Report 156   1996

Infiltration drainage – Manual of good practice

Roger Bettess BSc PhD MCIWEM

CONSTRUCTION INDUSTRY RESEARCH AND INFORMATION ASSOCIATION
6 Storey's Gate London SW1P 3AU
E-mail switchboard@ciria.org.uk
Tel: (0171) 222 8891   Fax: (0171) 222 1708

Summary

This manual provides a guide to good practice for those involved in the planning, appraisal, approval, funding, design, construction and maintenance of infiltration drainage systems who wish to use infiltration drainage as a method to control and dispose of stormwater. The manual discusses the advantages and disadvantages of such systems and provides the information to assist practitioners to decide whether, in given circumstances, infiltration techniques are appropriate. The manual also provides information which will enable its readers to:

(a) conduct field tests and relate the data to design
(b) design a range of types of infiltration systems.

Water quality is an important aspect of the design of infiltration systems and guidance on these issues and suitable pollution prevention measures is provided. The legal aspects are discussed as they apply to England and Wales.

The manual also constitutes NRA R&D Report 26 produced through NRA Project 333.

Roger Bettess BSc PhD MCIWEM
Infiltration drainage — Manual of good practice
Construction Industry Research and Information Association
CIRIA Report 156, 1996

© CIRIA 1996

ISBN 0 86017 457 3

ISSN 0305 408X

Keywords	
Urban drainage, stormwater, surface water sewer, runoff, infiltration drainage, soakaways, infiltration trenches, infiltration basins, infiltration blankets, swales, water quality, groundwater pollution, hydraulics, capital and current costs, law, geotechnical aspects.	

Reader interest	Classification	
Drainage engineers, house builders, developers, local authorities, water companies, sewerage undertakers, National Rivers Authority, water quality regulators, planning authorities, building control authorities.	AVAILABILITY	Unrestricted
	CONTENT	Review/advice/guidance
	STATUS	Committee guided
	USER	Developers and drainage engineers

Published by CIRIA, 6 Storey's Gate, Westminster, London SW1P 3AU. All rights reserved. No part of this publication may be reproduced or transmitted in any form or by any means, including photocopying and recording, without the written permission of the copyright holder, application for which should be addressed to the publisher. Such written permission must also be obtained before any part of this publication is stored in a retrieval system of any nature.

Foreword

Infiltration drainage systems may be used to dispose of surface water runoff from urban and highway areas by recharge into the ground. These systems allow stormwater to infiltrate into the soil over a period of time and also provide some detention storage during the storm event. Examples of such infiltration systems include individual and linked soakaways, infiltration trenches, infiltration basins, swales, infiltration pavements and infiltration blankets.

Infiltration should be seen as one of a number of methods of controlling surface water runoff. The use of infiltration drainage systems reduces the quantity of the water that has to be disposed of through surface water drains or sewers to local watercourses or treatment works. It may be especially useful in the on-site drainage of small new developments which would otherwise need to have new surface water sewers built to accept the additional runoff.
The re-direction of water out of a catchment area by pipe may reduce flows in water courses with the obvious loss of amenity and drinking water for animals and may also reduce aquifer re-charge.

There are situations, however, in which the use of infiltration will not be appropriate. This may be because the nature of the ground does not allow sufficient infiltration, the quality of the water infiltrating may pose a threat to groundwater resources, geotechnical problems may be too severe or the natural water table may be too close to the surface.

Until recently the most common form of infiltration drainage system used in the UK was the soakaway. This was normally used to provide drainage for a single house or small development. Lately soakaways have been used for a wider range of applications, some of which involve larger and more sophisticated structures than those used in the past. This has required a greater degree of understanding in the design of such systems. Despite their widespread use outside the UK, however, infiltration systems other than soakaways have not yet been so widely used here.

CIRIA reports R123 and R124 on the 'Scope for control of urban runoff' (CIRIA 1992a and 1992b), identified the potential for and benefits of using infiltration systems to reduce/attenuate storm flows before they enter piped drainage systems. It was in the context of a growing awareness of the potential importance of infiltration systems that this later project was conceived.

This manual provides a guide to good practice for those involved in the planning, appraisal, approval, funding, design, construction and maintenance of infiltration drainage systems who wish to use infiltration drainage as a method to control and dispose of stormwater. The manual discusses the advantages and disadvantages of such systems and provides the information to assist practitioners to decide whether, in given circumstances, infiltration techniques are appropriate. The manual also provides information which will enable its readers to:

(a) conduct field tests and relate the data to design
(b) design a range of types of infiltration systems.

Water quality is an important aspect of the design of infiltration systems and guidance on these issues and suitable pollution prevention measures is provided. The legal aspects are discussed as they apply to England and Wales.

The manual also constitutes NRA R&D Report 26 produced through NRA Project 333.

This manual does not contain all the detailed information which has been obtained during the course of the project and which forms the background to its recommendations. Reference may be made to the following reports which are published separately in support of the manual:

Project Report 21 Infiltration drainage − Literature review

This report identifies published information relevant to the manual, collates references, reviews key publications, and comments on the development of practice and knowledge identified in the literature.

This report also constitutes NRA R&D Note 484 produced through NRA Project 333.

Project Report 22 Infiltration drainage − Case studies of UK practice

The report presents a summary of the main types of infiltration system used in the UK, based on an inspection of these systems and the experiences gained in their use. Thirteen case studies are presented under standard heads: title, location, client, date of construction, design, contact, description, design method, construction, maintenance/present performance, design/maintenance, and lessons for future practice.

This report also constitutes NRA R&D Note 485 produced through NRA Project 333.

Project Report 23 Infiltration drainage − Hydraulic design

The report describes a theoretical study of the flow from a soakage pit and an analytical approach based on this. Procedures are developed and described which allow the designer to conduct field tests and to dimension infiltration systems of various types.

This report also constitutes NRA R&D Note 486 produced through NRA Project 333.

Project Report 24 Infiltration drainage − Appraisal of costs

The report identifies the costs and benefits pertinent to the economic and financial analysis of alternative drainage schemes, develops procedures for estimating the capital and recurrent costs, and shows how economic and financial cost comparisons may be made.

This report also constitutes NRA R&D Note 487 produced through NRA Project 333.

Project Report 25 Infiltration drainage − Legal aspects

The report presents a balanced and comprehensive coverage of the key (but often complex) legal issues involved in urban drainage as they apply to England and Wales and summarises these at a suitable level of detail for readers seeking an overview of the subject.

This report also constitutes NRA R&D Note 488 produced through NRA Project 333.

Further information on the water quality aspects of drainage systems can be found in:

* *Design of flood storage reservoirs*
  M.J. Hall, D.L. Hockin and J.B. Ellis
  CIRIA Book 14, 1993

* *Control of pollution from highway drainage discharges*
  CIRIA Report 142, 1994

* *Design and management of constructed wetlands for the treatment of wastewater*
  CIRIA Funders Report FR/CP/34, 1996

* *Use of industrial by-products in road construction: Water quality effects*
  CIRIA Funders Report FR/IP/11, 1996

Acknowledgements

The manual and the supporting reports were produced as a result of CIRIA Research Project 448. The objective of the project was to produce a manual of good practice for the design, construction and maintenance of infiltration systems for the on-site control and disposal of stormwater runoff from small-scale residential or commercial development upstream of an area with existing sewers.

The manual was compiled from information obtained from studies of various aspects of infiltration drainage. In addition to this report, these included: a literature survey; a survey of current practice; case studies of existing systems in the UK; a study of water quality aspects and methods for the control of pollution; development of hydraulic analysis and design methods; a study of geotechnical aspects; appraisal of costs; and a study of the legal implications of infiltration drainage. Reports are available from CIRIA on each of these aspects, apart from the work on a survey of current practice, a study of water quality aspects and a study of geotechnical aspects which are available from HR Wallingford in the form of project records.

The project was developed jointly by CIRIA and HR Wallingford. CIRIA had completed the 'Scope for control of urban run-off' project (published in 1992 as Reports 123 and 124) which among other matters had identified the potential for and benefits of the more widespread use of infiltration techniques for stormwater control and disposal. HR Wallingford had completed a research project on the hydraulic performance of soakaways which modelled the dispersal of water from cylindrical soakaways and determined the effects of scale between field tests and prototype performance. A proposal was prepared and submitted jointly by CIRIA and HR Wallingford to the Department of the Environment.

The work was carried out by HR Wallingford under contract to CIRIA in the period October 1991 to March 1995. The various studies were carried out either by HR staff or by other persons or organisations under subcontract to HR Wallingford. The names of those responsible for these studies and preparing the reports are listed below.

The Project Steering Group which guided the work at HR Wallingford was representative of a broad range of interested organisations. The names of its members and those who contributed financially to the work are listed below.

HR's Project Supervisor was Dr W R White and HR's Project Managers were Mr David Watkins (October 1991 – September 1992); Ms Amanda Davis (September 1992 – July 1993); Dr Roger Bettess (July 1993 – March 1996).

The studies and reports were prepared by:

Manual of good practice Report 156	Dr Roger Bettess – HR Wallingford
Literature review Project Report 21	Mr David Watkins – University of Exeter*
Case studies of UK practice Project Report 22	Professor Chris Pratt – Coventry University
Hydraulic design Project Report 23	Dr Roger Bettess – HR Wallingford Ms Amanda Davis* Mr David Watkins – University of Exeter*

Appraisal of costs Project Report 24	Mr John Roberts – Ove Arup & Partners
Legal aspects Project Report 25	Professor William Howarth – University of Kent at Canterbury Mr Alastair Brierley – Garrard Mitchell & Co.
Survey of practice (Project record)**	Mr David Watkins – University of Exeter* Professor Chris Pratt – Coventry University
Geotechnical aspects (Project record)**	Professor Tom Hanna – Consultant
Water quality aspects (Project record)**	Mrs Yvonne Walden*

formerly HR Wallingford
**available from HR Wallingford*

Project Steering Group

CIRIA and HR Wallingford wish to express their appreciation to the members of the Project Steering Group which guided the work and agreed the text of the manual and reports. Excluding those involved with the work as research contractors, the Project Steering Group comprised:*

Mr B H Rofe (Chairman)	Rofe Kennard & Lapworth
Mr R Addison	Anglian Water
Mr P J Allen	Department of Transport (from February 1993 to July 1993)
Mr D Askew	Northumbrian Water
Mr I Bernard	Thames Water
Mr R E Boots	Department of Transport (from September 1993)
Mr G Brennan	Concrete Pipe Association
Mr R Cook	Hydro Research and Development Ltd
Mr M J H Davis	formerly Tewkesbury Borough Council (from February 1993)
Mr R Farley	Severn-Trent Water
Mr P Freckleton	Yorkshire Water (from May 1992)
Mr D Homer	Derek Homer Associates
Mr P E Johnson	Department of Transport (until February 1993)
Mr P Myerscough	Yorkshire Water (until February 1992)
Dr A T Newman	NRA South Western Region
Mr J J M Powell	Building Research Establishment
Mr N J Price	TBV Consult (successor to PSA Specialist Services)
Mr A Proctor	Welsh Water (until July 1993)
Mr C R Rees	Albion Concrete Products Ltd
Mr R B Rosbrook	Southern Water Services (until February 1993)
Dr H R Thomas	University of Wales College of Cardiff
Mr R Tyler	Wessex Water
Mr P B Woodhead	Department of the Environment
Dr J A Payne (Secretary)	CIRIA (to December 1992)
Dr D E Wright (Secretary)	CIRIA (from January 1993)

Corresponding members

Mr I Davis	National House Building Council
Mr B R Fossett	CIN Properties Limited
Mr M Price	University of Reading

Dr David Wright

Dr David Wright was the CIRIA Project Manager for this study from January 1993 until his untimely death in April 1995. The report was substantially complete before his death and owes much to the industry and vision which he brought to all his work. It is hoped that this work will act as a small tribute to him and his long devotion to the subject of stormwater drainage and control.

---

* affiliations correct at February 1994. Dates in brackets refer to membership of PSG.

Financial contributions

The project was funded entirely by special contributions. CIRIA and HR Wallingford wish to express their gratitude to the following organisations:

Government/agency

Department of the Environment
Department of Transport
 (*via* Transport Research Laboratory)
National Rivers Authority

Industry

CIN Properties
Derek Homer Associates
National House Building Council
Concrete Pipe Association
Hydro Research and Development

Water companies

Anglian
Northumbrian
Severn-Trent
Southern
Thames
Welsh
Wessex
Yorkshire

As from April 1996, the functions of Her Majesty's Inspectorate of Pollution (HMIP), the National Rivers Authority (NRA) and the Waste Regulatory Authorities will be taken over, in England and Wales, by the Environment Agency. Its equivalent functions in Scotland will be taken over by the Scottish Environment Protection Agency and in Northern Ireland, by the Environment and National Heritage Agency.

Contents

Figures xi
Tables xii
Notation xiii
Guide to use xiv

1 INTRODUCTION 1
   1.1 What is infiltration drainage? 1
   1.2 The role of infiltration in stormwater drainage practice 1
   1.3 Background to project 1
   1.4 Purpose of manual 2
   1.5 Scope of manual 2
   1.6 Sources of information 2
   1.7 CIRIA'S project on Highway Drainage Systems 2
   1.8 Associated publications 3

2 INFILTRATION DRAINAGE SYSTEMS 4
   2.1 Infiltration drainage in the context of stormwater drainage 4
      2.1.1 Urban development of a catchment 4
      2.1.2 Alternative methods of runoff control 4
      2.1.3 Typical uses for infiltration drainage systems 4
      2.1.4 Basis for comparing methods of stormwater control and disposal . . . 5
   2.2 Principles of infiltration drainage 7
   2.3 Main types of infiltration systems 7
      2.3.1 Ground surface infiltration systems 7
      2.3.2 Sub-surface infiltration systems 11
   2.4 Usage of infiltration systems 14
   2.5 Advantages and disadvantages of infiltration 16

3 GUIDE TO THE DESIGN AND APPROVAL PROCEDURE 17
   3.1 Introduction 17
   3.2 Overall design and approval process 17
   3.3 Feasibility of using an infiltration system 17
   3.4 Outline approval 19
      3.4.1 Consent to discharge 19
      3.4.2 Maintenance or adoption by an authority 20
   3.5 Site investigations 21
      3.5.1 Infiltration tests 21
      3.5.2 Geotechnical investigation 21
   3.6 Selection of infiltration system 21
   3.7 Design of infiltration system 22
      3.7.1 Geotechnical investigation 28
      3.7.2 Hydraulic design 28
      3.7.3 Pollution control and mitigation 28
      3.7.4 Maintenance 28
      3.7.5 Appraisal of costs 31
   3.8 Approval from appropriate authorities 33
      3.8.1 Approval from the National Rivers Authority 33
      3.8.2 Maintenance or adoption 33
      3.8.3 Planning permission 34
      3.8.4 Building control 35

4 GENERAL DESIGN CONSIDERATIONS ... 36
- 4.1 Introduction ... 36
- 4.2 Appraisal of costs ... 36
  - 4.2.1 Introduction ... 36
  - 4.2.2 Perspective on costs ... 37
  - 4.2.3 Costing of infiltration systems ... 37
  - 4.2.4 Cost appraisal ... 39
  - 4.2.5 Example of cost appraisal ... 40
- 4.3 Hydrological and hydrogeological principles of infiltration ... 43
  - 4.3.1 Hydrological processes ... 43
  - 4.3.2 Enhanced infiltration for stormwater disposal ... 44
  - 4.3.3 Infiltration capacity of soils ... 44
  - 4.3.4 So where does the water go? ... 45
  - 4.3.5 Site determination of infiltration coefficient ... 46
- 4.4 Hydraulic design ... 47
  - 4.4.1 Determination of design rainfall events ... 47
  - 4.4.2 Hydraulic design of infiltration system ... 50
- 4.5 Geotechnics ... 61
  - 4.5.1 Introduction ... 61
  - 4.5.2 Data collection ... 61
  - 4.5.3 Geotechnical issues ... 61
  - 4.5.4 Checklist for design of infiltration systems ... 64
- 4.6 Pollution control requirements and methods of mitigation ... 64
  - 4.6.1 Sources of pollutants ... 65
  - 4.6.2 Removal of pollutants ... 65
  - 4.6.3 Groundwater protection ... 69
  - 4.6.4 A checklist for water quality ... 71
- 4.7 Maintenance responsibilities ... 72

5 DESIGN CONSIDERATIONS FOR PARTICULAR SYSTEMS ... 74
- 5.1 Introduction ... 74
- 5.2 Infiltration basins and swales ... 74
  - 5.2.1 Description ... 74
  - 5.2.2 Suitability of site ... 75
  - 5.2.3 Hydraulic design ... 76
  - 5.2.4 Construction ... 76
  - 5.2.5 Selection of vegetation type ... 80
  - 5.2.6 Maintenance ... 81
  - 5.2.7 Advantages and disadvantages ... 82
- 5.3 Infiltration trenches with surface inflow ... 82
  - 5.3.1 Description ... 82
  - 5.3.2 Suitability of site ... 82
  - 5.3.3 Hydraulic design ... 82
  - 5.3.4 Construction ... 84
  - 5.3.5 Maintenance ... 84
  - 5.3.6 Advantages and disadvantages ... 85
- 5.4 Infiltration pavements ... 85
  - 5.4.1 Description ... 85
  - 5.4.2 Suitability of site ... 85
  - 5.4.3 Hydraulic design ... 86
  - 5.4.4 Construction ... 87
  - 5.4.5 Maintenance ... 87
  - 5.4.6 Advantages and disadvantages ... 88
- 5.5 Infiltration blankets ... 88
  - 5.5.1 Description ... 88
  - 5.5.2 Suitability of site ... 88
  - 5.5.3 Hydraulic design ... 88
  - 5.5.4 Construction ... 88

		5.5.5	Maintenance	90
		5.5.6	Advantages and disadvantages	91
	5.6	Soakaways including circular, trench and linked soakaways		91
		5.6.1	Description	91
		5.6.2	Suitability of site	91
		5.6.3	Hydraulic design	91
		5.6.4	Construction	93
		5.6.5	Maintenance	94
		5.6.6	Advantages and disadvantages	94
6	LEGAL AND ADMINISTRATIVE ASPECTS OF INFILTRATION DRAINAGE			95
	6.1	Summary		95
	6.2	Legal aspects		95
		6.2.1	Pollution of groundwater	95
		6.2.2	Protection of groundwater resources against pollution	95
		6.2.3	Adoption by sewerage undertakers	96
		6.2.4	Acceptance of responsibility by highway authorities	96
		6.2.5	Adoption by land drainage bodies	96
		6.2.6	Maintenance responsibilities	97
		6.2.7	Civil liability and redress	97
		6.2.8	The European Community Groundwater Directive	98
	6.3	Administrative aspects		98
		6.3.1	Roles of the local authority	98
		6.3.2	Planning controls	99
		6.3.3	Building control	100
		6.3.4	Consent to discharge and land drainage consent	100
7	FURTHER ISSUES			101
	7.1	Legislation		101
	7.2	Adoption		101
	7.3	Long-term performance and maintenance		101
	7.4	Maintenance procedures		101
	7.5	Water quality		102
	7.6	Analysis of the risk of groundwater pollution		102
	7.7	Cost data		102
8	CONCLUSIONS			103

References 104

Appendix Relationship between the method of hydraulic calculations in CIRIA Report 124 (Volume 3), BRE Digest 365 and the method presented in this manual 106

Figures

Figure 2.1	Options for the control of urban runoff	5
Figure 2.2	Urban runoff and the catchment	6
Figure 2.3	Infiltration pavement	8
Figure 2.4	Infiltration basin	9
Figure 2.5	Swale	10
Figure 2.6	Soakaway	12
Figure 2.7	Trench soakaway	13
Figure 2.8	Infiltration trench with surface inflow	14
Figure 2.9	Infiltration blanket	15

Figure 3.1	Flowchart A: Overall procedure for designing and obtaining approval for an infiltration system	18
Figure 3.2	Flowchart B: Checklist of questions at feasibility stage	19
Figure 3.3	Flowchart C: Outline approvals	20
Figure 3.4	Flowchart D: Site investigation	21
Figure 3.5	Flowchart E: Selection of type of infiltration system (sub-surface or ground)	23
Figure 3.6	Flowchart F: Selection of type of sub-surface infiltration system	24
Figure 3.7	Flowchart G: Selection of type of ground infiltration system	25
Figure 3.8	Flowchart H: Overall design procedure	26
Figure 3.9	Flowchart I: Geotechnical assessment	27
Figure 3.10	Flowchart J: Hydraulic design of infiltration systems	29
Figure 3.11	Flowchart K: Pollution control and mitigation	30
Figure 3.12	Flowchart L: Maintenance and adoption	31
Figure 3.13	Flowchart M: Appraisal of costs	32
Figure 3.14	Flowchart N: Obtaining approval for an infiltration system	34
Figure 4.1	Infiltration system used for stormwater runoff control and disposal	44
Figure 4.2	(a) Rainfall hyetograph (b) Hydrograph of inflow into infiltration system	48
Figure 4.3	Values of rainfall ratio r in UK	49
Figure 4.4	Detailed flow chart for hydraulic design	52
Figure 4.5	Graph to determine maximum depth for 3-D infiltration systems	58
Figure 4.6	Examples of simple oil and petrol interceptor designs	67
Figure 4.7	Example of a highway runoff infiltration system with an interceptor in place (after Price, 1992)	68
Figure 4.8	Typical arrangement for a reed-bed treatment system	69
Figure 5.1	Lattice of blocks with infiltration surface set below load bearing surface	87

Tables

Table 4.1	Indicative costs of maintenance of drainage systems	39
Table 4.2	Example – Comparison of capital costs of positive and infiltration drainage systems	41
Table 4.3	Example – Comparative present values of drainage system options	42
Table 4.4	Typical infiltration coefficients based on soil texture (Watkins, 1995)	45
Table 4.5	M10 rainfall intensity (mm/h) for duration D and ratio r (England & Wales)	51
Table 4.6	Factor of safety, F, for use in hydraulic design	53
Table 4.7	Typical values for the porosity of fill material	54
Table 4.8	Reported ranges of pollutant levels in runoff from roads found in various locations (Colwill et al 1984; Strecker et al 1990).	66
Table 4.9	NRA Acceptability Matrix for Discharges to Underground Strata (from NRA Groundwater Protection Policy, 1992)	70
Table 5.1	Assessment of soil potential for plant growth, (from Coppin and Richards (1990))	76
Table 5.2	Rainfall duration and intensity	77
Table 5.3	Values of rainfall intensity, duration, b, a and h_{max}	78
Table 5.4	Values of rainfall intensity, duration, b, a and h_{max}	78
Table 5.5	Values of rainfall intensity, duration, b, a and h_{max} for an infiltration trench	83
Table 5.6	Values of rainfall intensity, duration and highest water level, h_{max}, for an infiltration pavement	86
Table 5.7	Values of rainfall intensity, duration and highest water level, h_{max}, for an infiltration blanket	89
Table 5.8	Values of rainfall intensity, duration and infiltration area, A_b, for an infiltration blanket	90
Table 5.9	Values of rainfall intensity, duration, b, a and h_{max} for a cylindrical soakaway	92

Notation

A_b	area of base of infiltration system
A_D	area to be drained
a	parameter in equation for 3-D infiltration
a_{p50}	area of base and sides of test pit at 50% of the depth
b	parameter in equation for 3-D infiltration
D	duration of storm event
exp	exponential function
F	factor of safety
h_{max}	maximum depth of water in infiltration system during a storm
i	rainfall intensity
L	length of excavation for a soakaway
$\log_e$	natural logarithm
M	depth of rainfall (in mm)
$MT\text{-}D$	rainfall event with depth of rainfall of M mm, a return period of T years and a duration of D (minutes or hours as defined)
n	porosity of fill in infiltration systems (voids volume/total volume)
P	perimeter of vertical-sided infiltration system
PV	present value of a future cash flow
Q	inflow into infiltration system
q	infiltration coefficient of soil
R	ratio of the drained area to the infiltration area
r	ratio of depths of rainfall of given return period, but different durations $\left(\dfrac{MT\text{-}D_1}{MT\text{-}D_2}\right)$
r'	radius of ring sections in a soakaway constructed from perforated concrete rings
T	return period of storm event
$t_{p75\text{-}25}$	time for test pit to empty from 75% full to 25% full
$V_{p75\text{-}25}$	volume of test pit between 75% full and 25% full
W	width of excavation for a soakaway

Guide to use

The manual is designed for use by professional staff of the water utilities, sewerage undertakers, NRA, developers, local authorities, highway authorities, planners, architects and other organisations involved in the surface water drainage of new development. The manual is structured as follows:

Section 1 introduces the topic of infiltration systems and presents the background to the present study. It describes the scope of the manual, the major sources of information used in the project and relationship of the present work with the related CIRIA project on highway drainage (CIRIA, 1994).

Section 2 discusses the role of infiltration drainage in the context of stormwater drainage as a whole and describes in outline the main types of infiltration systems. It also describes the advantages and disadvantages of using infiltration to control stormwater runoff.

Section 3 provides a framework for designing and obtaining approval for an infiltration system. The section outlines a logical approach to the use of the manual and at each stage directs the reader to the appropriate subsection.

Section 4 covers a number of general issues which are common to all types of infiltration systems, such as appraisal of costs, hydrology and hydrogeology, methods for hydraulic design, geotechnics, pollution control and maintenance.

Section 5 deals with design issues specific to particular types of infiltration systems and thus has separate subsections devoted to each type of system. These should be read in conjunction with the more general text in Section 4.

Section 6 describes the legal and administrative aspects of infiltration drainage and provides an overview of planning, building control, NRA and other regulatory influences as they apply in England and Wales.

Section 7 discusses a number of issues which require clarification or further investigation. These became apparent during the course of the study but were beyond its scope.

Section 8 summarises the main conclusions from the project.

As from April 1996, the functions of Her Majesty's Inspectorate of Pollution (HMIP), the National Rivers Authority (NRA) and the Waste Regulatory Authorities will be taken over, in England and Wales, by the Environment Agency. Its equivalent functions in Scotland was be taken over by the Scottish Environment Protection Agency and in Northern Ireland, by the Environment and National Heritage Agency.

1 Introduction

1.1 WHAT IS INFILTRATION DRAINAGE?

Infiltration drainage systems may be used to dispose of stormwater from urban and highway areas by recharge into the ground. These systems allow stormwater to infiltrate into the ground over a period of time and also provide some detention storage during the storm event, depending on type. Examples of such infiltration systems include individual and linked soakaways, infiltration trenches, infiltration basins, swales, infiltration blankets and infiltration pavements.

1.2 THE ROLE OF INFILTRATION IN STORMWATER DRAINAGE PRACTICE

Infiltration should be seen as one of a number of methods of controlling stormwater runoff. The use of infiltration drainage systems reduces the quantity of water that has to be disposed of through surface water drains or sewers to local watercourses or treatment works. It can be especially useful in the on-site drainage of small new developments which would otherwise need to have new stormwater sewers built to accept the additional run-off.

There are situations in which the use of infiltration will not be appropriate. For example, this may be because of the low infiltration capacity of the ground, the quality of the water infiltrating may pose a threat to groundwater resources, or the natural water table may be too close to the surface. This is discussed in greater detail in Section 2.

1.3 BACKGROUND TO PROJECT

Until recently the most common form of infiltration drainage system was the soakaway. This was normally used to provide drainage for a single house or small development. Design advice was contained in BRE Digest 151 and British Standard BS 8301. In the last few years soakaways have been used for a wider range of applications, some involving larger and more sophisticated structures than those used previously. This has required a greater degree of understanding in the design of such systems as provided by BRE Digest 365 (1991).

The CIRIA project on the 'Scope for control of urban runoff', (CIRIA 1992a and 1992b) identified the potential for and benefits of using infiltration systems to reduce/attenuate storm flows before they enter piped drainage systems. Despite their widespread use outside the UK, however, infiltration systems other than soakaways have not yet been so widely used in the UK.

Alongside these developments an increasing knowledge of the performance of infiltration systems has slowly accumulated. This has in part comprised practical experience of infiltration systems, some of which have been constructed in the UK but many of which have been constructed overseas. In parallel with this practical experience, studies have been carried out into technical aspects related to infiltration, an example of which is the DOE-funded study carried out by HR Wallingford into the hydraulics of soakaways (Watkins, 1991). Much of this knowledge is not readily accessible to the practitioner.

It was in the context of this growing awareness of the potential importance of infiltration systems and increasing knowledge that the present project was conceived.

1.4 PURPOSE OF MANUAL

The manual provides a guide to good practice for those involved in the approval, funding, design, construction and maintenance of infiltration drainage systems who wish to use infiltration drainage as an on-site method to control and dispose of stormwater. The manual discusses the advantages and disadvantages of such systems and provides the information to enable practitioners to decide whether, in given circumstances, infiltration techniques are appropriate. The manual also provides information which will enable its readers to:

(a) study the feasibility and select the type of infiltration system
(b) conduct field tests and relate the data to design
(c) design a chosen type of infiltration system
(d) incorporate suitable pollution prevention measures into the design.

1.5 SCOPE OF MANUAL

The manual covers the planning, design, economic and financial appraisal, construction and maintenance of a range of infiltration systems. Water quality is an important aspect of the design of infiltration systems and guidance on this issue is provided. The legal aspects and information on obtaining approval for an infiltration system from the appropriate authorities are discussed as they apply to England and Wales.

1.6 SOURCES OF INFORMATION

The manual has been compiled from information contained in a number of detailed reports on studies of different aspects of infiltration systems. These studies have included:

- a review of the literature concerned with infiltration drainage systems

- a summary of the characteristics and performance of a variety of infiltration drainage systems by the use of case studies of UK practice

- a literature review on water quality aspects including pollutant source and control methods

- development of hydraulic design methods for infiltration systems

- an appraisal of costs

- a review of the legal considerations

- a survey of current practice

- a study of the geotechnical aspects of infiltration drainage.

1.7 CIRIA'S PROJECT ON HIGHWAY DRAINAGE SYSTEMS

At the same time as the work on the infiltration manual and reports was proceeding, CIRIA Project RP473 was looking at the control of pollution from highway drainage systems as it might affect both groundwater and surface waters. The purpose of RP473 was primarily to develop policy guidance, whereas the infiltration drainage project was mainly to provide design guidance in a manual format. Care has been taken to ensure that this manual and the report on RP 473, CIRIA (1994) are compatible in all essential respects. The two projects have relied on the same database and knowledge in respect of the risk to groundwater by pollution from surface runoff from highways and to that extent they overlapped.

Further information on the water quality aspects of drainage systems can be found in:

* *Design of flood storage reservoirs*
  M.J. Hall, D.L. Hockin and J.B. Ellis
  CIRIA Book 14, 1993

* *Control of pollution from highway drainage discharges*
  CIRIA Report 142, 1994

* *Design and management of constructed wetlands for the treatment of wastewater*
  CIRIA Funders Report FR/CP/34, 1996

* *Use of industrial by-products in road construction: Water quality effects*
  CIRIA Funders Report FR/IP/11, 1996

1.8 ASSOCIATED PUBLICATIONS

The manual is intended to be a stand-alone document which contains all the information essential for those wishing to plan, design, construct and use infiltration drainage systems. As a manual, however, it does not contain all the background and detailed information which has been obtained during the course of the project. CIRIA is, therefore, publishing the following reports as a companion set:

Project Report 21 *Infiltration drainage – Literature review*
Project Report 22 *Infiltration drainage – Case studies of UK practice*
Project Report 23 *Infiltration drainage – Hydraulic design*
Project Report 24 *Infiltration drainage – Appraisal of costs*
Project Report 25 *Infiltration drainage – Legal aspects*

The reports on a survey of current practice, water quality practice and geotechnical aspects are held as project records and are available for reference at HR Wallingford, Oxfordshire.

2 Infiltration drainage systems

2.1 INFILTRATION DRAINAGE IN THE CONTEXT OF STORMWATER DRAINAGE

2.1.1 Urban development of a catchment

The urban development of a catchment can have the following major effects on the hydrological regime:

- increased volumes of stormwater runoff
- higher peak flow rates and flood water levels
- lower base flows in water courses
- reduction of available storage in and conveyance capacity of river valleys
- reduction in soil moisture recharge leading to a reduction of groundwater resources
- increase in pollutant loads carried into sewers or surface waters.

The objectives of urban runoff control are to limit the quantity, location and frequency of flooding to an acceptable level and to maintain natural and artificial watercourses and surface water sources in a fit state for their other functions.

2.1.2 Alternative methods of runoff control

Figure 2.1 illustrates the main options which are available for control of storm runoff. There are four broad categories:

1. a reduction of the flow entering the drainage system
2. an attenuation of the flows prior to them entering the sewers or watercourse
3. an attenuation of the flows in the drainage system
4. an increase in the capacity of the drainage system.

As will be seen, infiltration drainage is part of the first category of stormwater control. The figure makes clear that this technique is one of a number of methods which are available. The role of infiltration systems is to reduce or eliminate the stormwater passing to conventional drainage systems and thus to treatment works or local watercourses.

2.1.3 Typical uses for infiltration drainage systems

As indicated by the schematic drainage layout shown in Figure 2.2, which shows the common routes for stormwater to follow between a development site and the river, infiltration methods may be considered as particularly appropriate for the on-site drainage of small-scale (i.e. up to 10 ha) residential, commercial or leisure development. In the absence of such methods, the runoff would pass undiminished into a downstream area and possibly cause problems there for the existing drainage system, requiring the use of additional sewer capacity or detention storage.

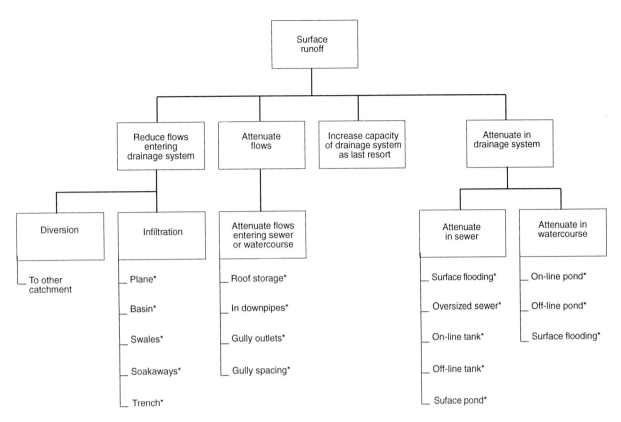

Options for use on the development sites

Figure 2.1 Options for the control of urban runoff

Infiltration techniques should, therefore, be seen as an alternative to providing new or upgraded storm sewerage systems within and downstream of the development, either on their own or in conjunction with conventional sewerage. Provided the design takes full account of the physical site conditions and, for example, the maintenance and pollutant control requirements, infiltration systems can be an attractive and cost-effective solution to many site drainage problems.

2.1.4 Basis for comparing methods of stormwater control and disposal

A proper cost comparison of an infiltration drainage system with the conventional sewerage alternative will take account of the following items:

(a) Infiltration option:

- infiltration drainage system (with any pipes integral to the system) for the development, including any pollution alleviation methods required
- any sewered connections which may still be required to existing sewers downstream
- any enhancements which may still need to be made to the existing sewers downstream
- benefits of groundwater recharge.

(b) Conventional option:

- sewerage for the development itself, including connections to existing sewers downstream
- enhancements to existing sewers, including detention storage and/or larger sewers
- any costs of disruption because of new sewer construction.

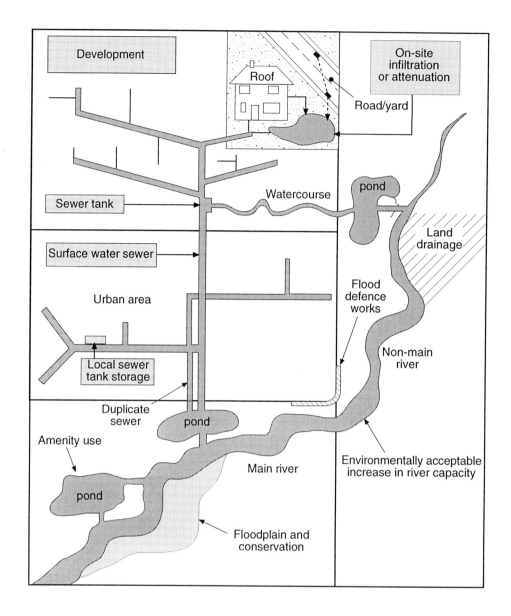

Figure 2.2 Urban runoff and the catchment

These procedures are explained in more detail in Section 4.2 (and CIRIA Project Report 24). The adoption of the infiltration option which reduces the flow entering the downstream drainage system can be most attractive.

Existing downstream sewers may have spare capacity. However, where this capacity was not specifically provided for the development in question, the costing of the conventional option should not make use of this spare capacity (in order to make a proper comparison with the infiltration option which would leave the spare capacity for other developments).

2.2 PRINCIPLES OF INFILTRATION DRAINAGE

Infiltration drainage systems dispose of urban stormwater by providing detention storage during the storm event and allowing the stormwater to infiltrate into the soil over a period of time. The importance of storage varies with the type of system. It is desirable that infiltration takes place into the unsaturated zone above the groundwater table. Where the discharge is below the groundwater table it is more commonly thought of as direct recharge to the groundwater. For a system to be effective it must have sufficient volume to store the runoff and have sufficient surface area in contact with the soil to allow infiltration of stormwater runoff. The size required depends on the hydraulic properties of the soil, the area which the system is draining and the chosen design rainfall events.

2.3 MAIN TYPES OF INFILTRATION SYSTEMS

Since infiltration rates into the ground are normally less than the rate of stormwater inflow, all infiltration systems involve an element of storage. Infiltration systems can be divided into two broad categories depending upon whether storage takes place *above* or *below* the ground surface. These are outlined below.

2.3.1 Ground surface infiltration systems

(a) General

Ground surface infiltration systems can utilise natural or artificial surfaces ranging from grass to permeable pavements. The shape of these systems may vary from near horizontal surfaces to basins or swales with distinct sides. Though there is a continuum of shapes they are conveniently divided into plane infiltration systems and basin infiltration systems.

(b) **Plane infiltration systems**

Infiltration pavements

Plane infiltration systems are represented by infiltration or permeable pavements, (see Figure 2.3). The surface of such systems may range from grass through permeable macadam to cellular concrete blocks. These systems are predominantly flat and their shape is commonly such that there is little storage per unit of infiltration area. They are frequently used to dispose of just the rainfall falling into the surface itself, as for example in a car park. Sometimes they may also be used to dispose of water from the roofs of adjacent buildings, but in these cases the area of the roofs is normally small in comparison with the infiltration area. The type of surface should be selected to be in sympathy with any use that is to be made of the area.

(c) **Basin infiltration systems**

Infiltration basins

To increase the amount of storage available per unit of infiltration area over plane infiltration systems a basin can be used (see Figure 2.4). The increase in storage means that these can be used to dispose of water from areas which are many times larger than the basin itself. An infiltration basin is an area of land surrounded by a bank or berm, which detains stormwater until it has infiltrated through the base and sides of the basin. Infiltration basins are sometimes also referred to as dry retention ponds.

Swales

A swale is a grass-lined channel with shallow side-slopes which may be used both to convey and to infiltrate stormwater (see Figure 2.5). To increase the infiltration and detention capacity of swales they can be provided with low check dams across their width.

Top view

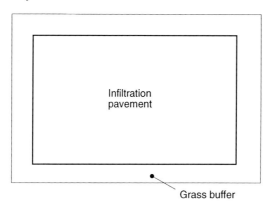

Side view

Figure 2.3 Infiltration pavement

Top view

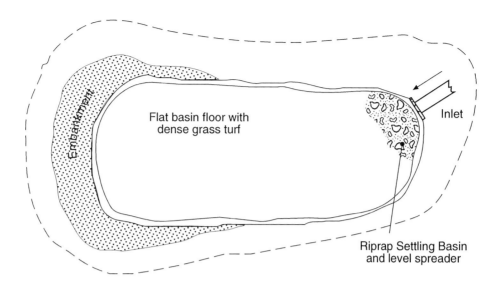

Side view

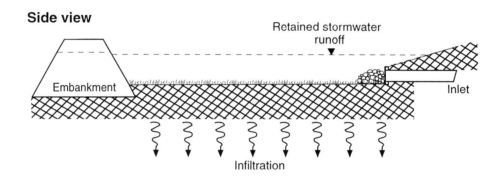

Figure 2.4 Infiltration basin

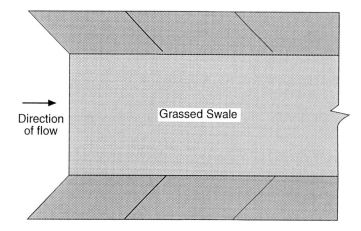

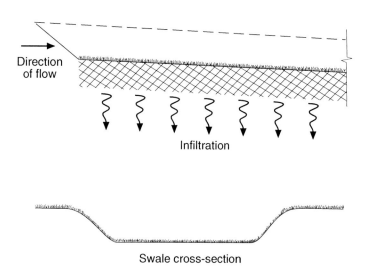

Figure 2.5 Swale

Dry retention ponds and swales can be included unobtrusively in a landscaped area. The necessity of using the area for water storage during a storm event, however, may limit alternative land use.

Infiltration basins and swales can improve the quality of the run-off. Deposition of particulate pollutants can occur together with the biological removal of some forms of pollutants. In well-designed systems the improvement in water quality can be substantial.

Polluted storm water can also be treated in wet detention ponds (see Section 4.6.2). As they are permanently wet, however, the base may silt up and form an effective seal, so that little or no infiltration occurs. Thus their hydraulic design is not covered by this manual. Since wet detention ponds can be used to control the quality of runoff, they can be used in conjunction with infiltration systems.

2.3.2 Sub-surface infiltration systems

In these systems the storage is provided below ground level. Such systems are thus unobtrusive and leave the overlying surface free for other uses. However, construction and, in particular, maintenance are normally more difficult than for surface systems.

Sub-surface systems are normally classified, according to the shape of the storage provided, as:

Soakaways

These may range from structures constructed, for example, of pre-cast, perforated concrete rings or loose-laid bricks to a simple rock-filled excavation. Traditionally, they have consisted of a cylindrical or rectangular hole excavated into the ground with a structure or stone-fill to maintain the shape of the excavation (see Figure 2.6). On large sites, individual soakaways can be linked together by pipes; for a given volume and storage, linked soakaways are likely to provide a greater infiltration area than an equivalent single soakaway.

Trench soakaways

For a given volume of excavation, better infiltration characteristics can be obtained from a soakaway in the form of a trench whose length is significantly longer than its width or depth (BRE Digest 365). These trenches are usually stone-filled. Such shapes of soakaway are normally called trench soakaways (see Figure 2.7).

Infiltration trenches with surface inflow

These are similar in form to trench soakaways, but instead of accepting runoff from a piped system they accept runoff through the surface. Such soakaways may, for example, be used along one edge of a car parking area to provide storm drainage (see Figure 2.8).

Infiltration blankets

These have many characteristics in common with infiltration pavements except that they are covered by soil or some other non-infiltrating surface. As the system is completely buried this allows alternative use of the ground surface. Stormwater is normally introduced into the blanket from one or more point sources (see Figure 2.9).

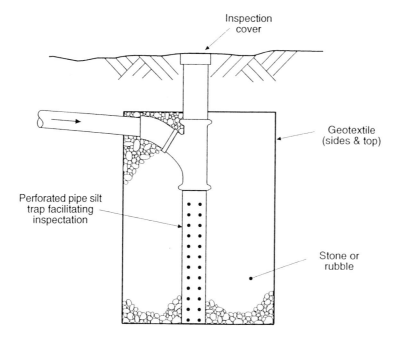

(a) Conventional soakaway details for a house

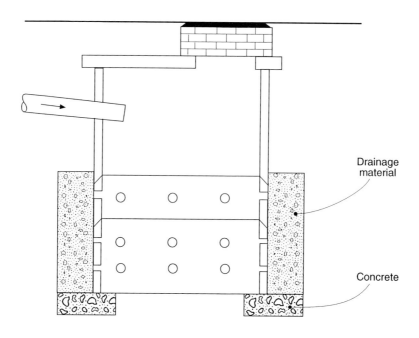

(b) Conventional estate development soakaway with rigid lining

Figure 2.6 Soakaway

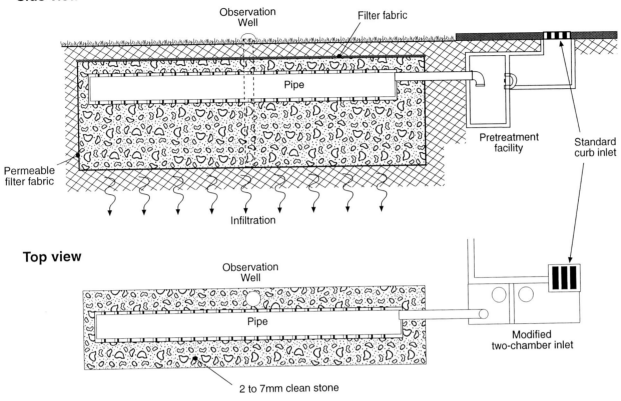

Figure 2.7 Trench soakaway

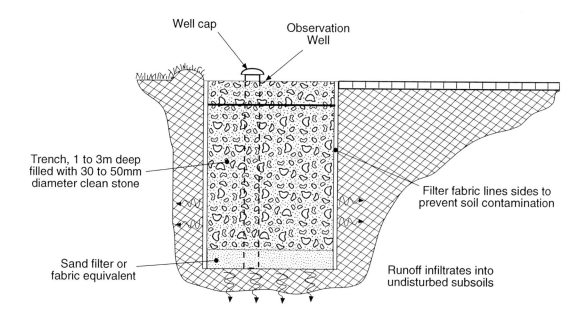

Figure 2.8 Infiltration trench with surface inflow

Overseas the term 'dry-well infiltration' is commonly used in place of soakaway.

2.4 USAGE OF INFILTRATION SYSTEMS

As part of the study a survey of current practice was carried out. In this section the findings of that survey are very briefly summarised.

Soakaways have been widely used throughout the country for a considerable period of time for the drainage of single houses or minor roads. The total number of soakaways constructed annually is believed to be well in excess of 65 000.

Top view

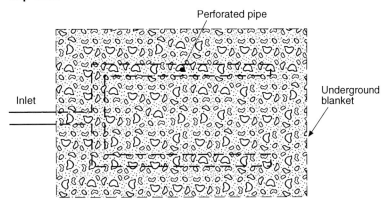

Side view

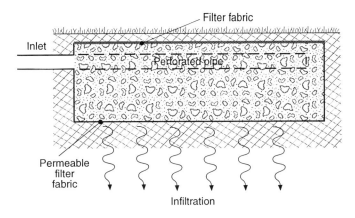

Figure 2.9 Infiltration blanket

Although the use of other types of infiltration systems has, until recently, been much less common in the UK, a growing range of types and sizes of infiltration systems are coming into use. Infiltration systems have been used to drain areas varying in size from a few square metres to many hectares. The areas drained have included car parks, roads, roofs, pavements, pedestrian areas, shopping developments, business parks and golf courses. While all respondents to the survey appeared to be familiar with soakaways, there was a significant percentage who were not aware of some of the other types of infiltration system. For example, only 42% of respondents were aware of the use of infiltration basins and only 56% were aware of the use of infiltration pavements. Even fewer respondents had been involved in the use of infiltration systems. For example, only 10% of respondents had been involved with infiltration pavements and only 15% with infiltration basins. It would appear that in some other countries (e.g. USA, Sweden) there is already extensive and successful experience of a range of infiltration systems, such as infiltration basins, swales and infiltration pavements. Their generally positive appreciation of the advantages of such systems should encourage their promotion in the UK (e.g. Pratt 1989).

It would appear, however, that, at the moment in the UK, the main reason for using infiltration systems is the unavailability of positive drainage, rather than a recognition of their advantages.

In the survey a variety of reasons for not using infiltration systems were given which were either technical, or legal in nature. The technical reasons included inappropriate ground conditions and the lack of suitable design methods. The legal concerns that have been expressed commonly relate to maintenance responsibility, adoption problems, and the liability for any pollution.

2.5 ADVANTAGES AND DISADVANTAGES OF INFILTRATION

The main advantages of infiltration systems are as follows:

- Infiltration reduces the quantity of water requiring conveyance in any piped system downstream of the development and decreases the flow in stormwater sewers and the risks of discharge from overflows in combined sewer systems.

- Infiltration may be used where there is no convenient existing storm drainage system to which connection can be made.

- By controlling stormwater close to source, infiltration drainage reduces the hydrological impact of urbanisation (as summarised in Section 2.1.1).

- Infiltration may be used where existing piped systems, or treatment works, are at capacity loading. It thus saves on the cost of enlarging the existing drainage system or providing additional detention storage with all the disruption and cost that this normally involves.

- Infiltration can be used to enhance recharge to groundwater in situations where the quality of stormwater runoff does not pose a threat to groundwater quality.

- Construction is normally simple and rapid.

- Whole-life costs may be less than for alternative systems.

The main disadvantages of infiltration systems include:

- The performance of infiltration systems depends on the properties of the soil in which they are constructed.

- Field tests are necessary in order to determine infiltration coefficients for design purposes.

- If the storm runoff is polluted, there is a risk that infiltration systems may introduce pollutants into the soil and ultimately into the groundwater. Various methods are available for reducing this risk. This is of particular concern when the system is used for drainage from industrial sites and highways and in such cases the use of infiltration may not be appropriate. Simple biological treatment methods may reduce levels of biodegradable pollutants but normally such methods are not adequate for non-biodegradable pollutants.

- The introduction of water into the soil may cause geotechnical problems.

- The adjacent soil can become blinded through ingress of silt and so infiltration systems require regular maintenance. Appropriate arrangements must be made for this to be carried out.

- It is the policy of the Statutory Sewerage Undertakers in England and Wales not to adopt infiltration systems.

- There may be a legal liability on the owner of the infiltration system for any pollution of the groundwater.

3 Guide to the design and approval procedure

3.1 INTRODUCTION

The overall procedure for planning, appraising, designing and obtaining approval for an infiltration system is described in this section.

The various stages in the process are described in detail in later sections of the manual and for more information on these the reader is referred to the appropriate parts of Sections 4, 5 and 6.

3.2 OVERALL DESIGN AND APPROVAL PROCESS

The overall process of design and approval is shown in Flowchart A, Figure 3.1. This shows the fundamental procedure to be followed which is broken down into more detailed actions in later flowcharts.

3.3 FEASIBILITY OF USING AN INFILTRATION SYSTEM

Before embarking on any detailed design and analysis, it is worth considering the feasibility of using infiltration systems to ensure that there is no fundamental reason why infiltration will not be appropriate. In general, at the feasibility stage a brief, qualitative assessment will be made, though for large systems it may become more extensive and even involve preliminary designs and costings. At this stage a number of issues should be examined. These will vary with the particular drainage problem under consideration: a brief checklist of the questions that should be considered is given below. This should cover the more commonly encountered issues.

Example checklist of questions at feasibility stage

- What is the volume of storm runoff that must be drained?

- Are there alternative, more economical methods for the disposal of storm drainage e.g. a nearby river or an existing sewer? If a sewer is available would its capacity need enhancing?

- What is the likely environmental and social impact of the proposed scheme?

- Does the quality of the runoff preclude the use of infiltration e.g. is the runoff from an industrial site and may it contain contaminating chemicals?

- Will the runoff quality and location preclude approval within the NRA Groundwater Protection Policy? To consider this, establish the relationship of the proposed site to NRA protection zones and assess the contaminant potential of the runoff.

- Would geotechnical problems associated with the soil type or the proximity of the infiltration system to slopes or the foundations of adjacent buildings exclude the possibility of infiltration e.g. is the soil made-up ground or swelling clay?

- Is the water table likely to approach the base of the infiltration system?

- What would be the consequences of flows in excess of the design flood?

This process is shown in Flowchart B, Figure 3.2.

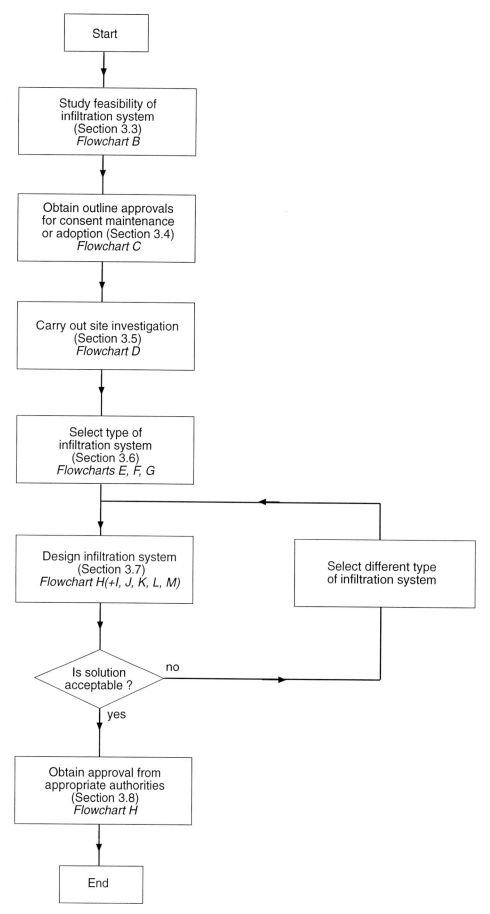

Figure 3.1 Flowchart A: Overall procedure for designing and obtaining approval for an infiltration system

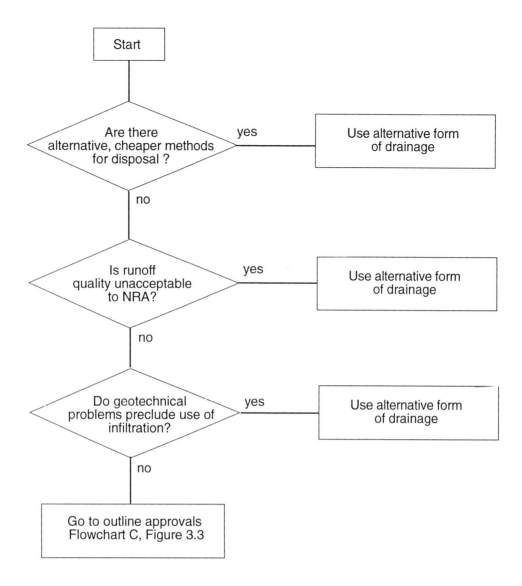

Figure 3.2 Flowchart B: Checklist of questions at feasibility stage

Once it has been established that an infiltration system is feasible the proposer/developer can proceed to outline approval.

3.4 OUTLINE APPROVAL

The process of obtaining outline approval is shown in Flowchart C, Figure 3.3.

3.4.1 Consent to discharge

The NRA operates a groundwater protection policy to conserve the quantity and quality of groundwater resources and sources and, depending upon the location of an infiltration system, there may be concern that pollution of an aquifer could occur.

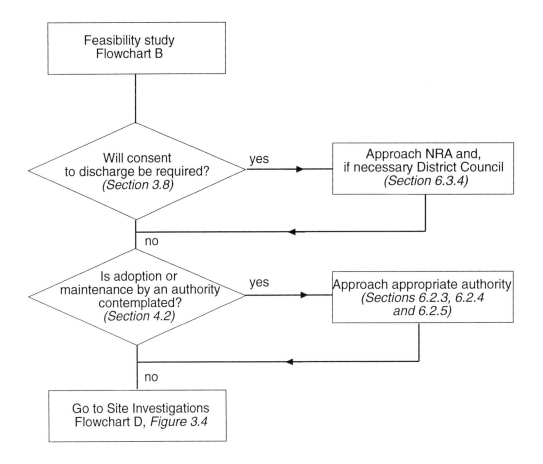

Figure 3.3 Flowchart C: Outline approvals

If the discharge from the infiltration system is of a polluting nature a 'Consent to Discharge' must be obtained. The NRA is the appropriate regulatory body to authorise a 'Consent to Discharge' and will require to know at least:

- the site of the proposed infiltration system
- the nature of the discharge, and in particular its likely quality
- the expected volumes to be discharged.

The NRA should be consulted at the earliest opportunity as final approval will normally require a more detailed design. See Section 3.8 for more details.

3.4.2 Maintenance or adoption by an authority

Some infiltration systems remain within the ownership and control of the site owner. In some circumstances, however, the owner may wish to pass the ownership or maintenance responsibility to some other authority. It is the policy of the Statutory Sewerage Undertakers in England and Wales not to adopt infiltration systems. If the system serves as a highway drainage system it may be open to adoption by a highway authority and in some cases local authorities may be prepared to adopt or maintain systems. If such an arrangement is being contemplated in return for a commuted sum for maintenance, the appropriate authority should be consulted at the earliest possible opportunity to establish its willingness or otherwise to adopt or maintain the system and any requirements it might have. See Sections 3.8 and 4.7 for more details.

If adoption of the system is proposed and agreed in principle, the adopting authority may impose their preferences on the type, size and method of operation of the infiltration system as a condition of adoption. (See final paragraph of Section 3.6).

3.5 SITE INVESTIGATIONS

The process of site investigation is shown in Flowchart D, Figure 3.4.

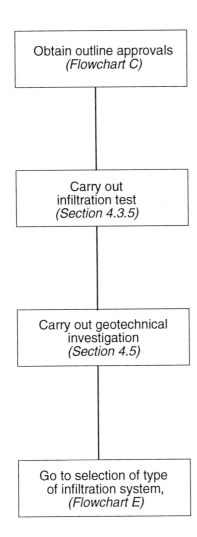

Figure 3.4 Flowchart D: Site investigation

3.5.1 Infiltration tests

An important factor in the design of an infiltration system is the rate at which water will infiltrate into the soil. As the infiltration capacity of soils varies, it is necessary to carry out an on-site infiltration test, the results of which are used in the hydraulic design. The detailed procedure for carrying out this test and analysing the results is described in Section 4.3.5.

3.5.2 Geotechnical investigation

Infiltration systems necessarily introduce water into the surrounding soil. The site should, therefore, be assessed for potential problems. See Sections 3.7.1 and 4.5 for more information.

3.6 SELECTION OF INFILTRATION SYSTEM

The process of selection of the infiltration system is shown in Flowcharts E, F and G, Figures 3.5 to 3.7.

The first major decision to be made in selecting an infiltration system is whether it should be a ground surface or a sub-surface system. The type of questions to consider are:

- Is excavation allowed on site?
- Is the area selected for the infiltration system to be used for other purposes?
- Are there any aesthetic reasons for having either a surface or a sub-surface system?
- Can water be stored temporarily on the surface?
- Is biological treatment required to mitigate pollution?
- What is the depth to the groundwater?

There are a large number of other factors that may influence the selection; these include:

- ease of maintenance
- location on site
- availability of land
- ease of construction
- planned sequence of construction.

A decision tree outlining these choices is shown in Flowchart E, Figure 3.5.

If a sub-surface system has been selected, the following type of questions should be considered in deciding (which type of system is appropriate):

- What is the requirement for storage per unit infiltration area?
- Is the drained area large?

A decision tree outlining the selection of sub-surface systems is given in Flowchart F, Figure 3.6.

If a ground surface infiltration system has been selected the following type of questions should be considered in deciding which type of system is appropriate :

- What is the requirement for storage per unit infiltration area?
- What storage volume will be required?

A decision tree outlining the selection of ground surface systems is given in Flowchart G, Figure 3.7.

If adoption of the system is proposed and agreed in principle, the adopting authority may impose their preferences on the type, size and method of operation of the infiltration system as a condition of adoption. In this case, these requirements would be a very strong factor in selecting the appropriate infiltration system for the development. If the developer subsequently believed that those conditions were imposing an unsuitable design or constraint on the proposals, then convincing evidence would need to be prepared to persuade the adopting authority to agree to alternative proposals.

3.7 DESIGN OF INFILTRATION SYSTEM

The process of design is shown in Flowchart H, Figure 3.8.

The design of the system needs to address a number of issues: it may prove necessary to carry out a number of iterations before a satisfactory design is achieved. The main areas to be considered are outlined in Flowcharts I, J, K, L and M (Figures 3.9 – 3.13) as noted below.

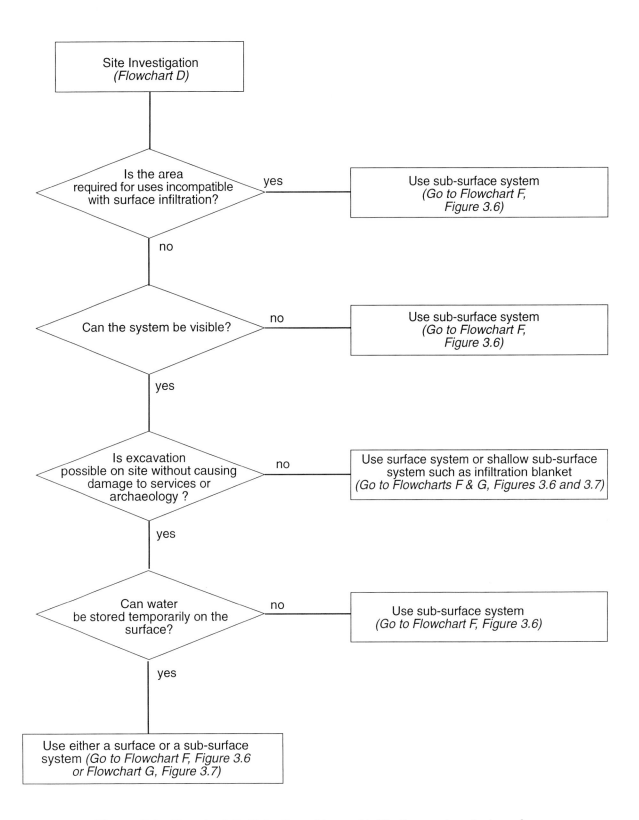

Figure 3.5 Flowchart E: Selection of type of infiltration system (sub-surface or ground)

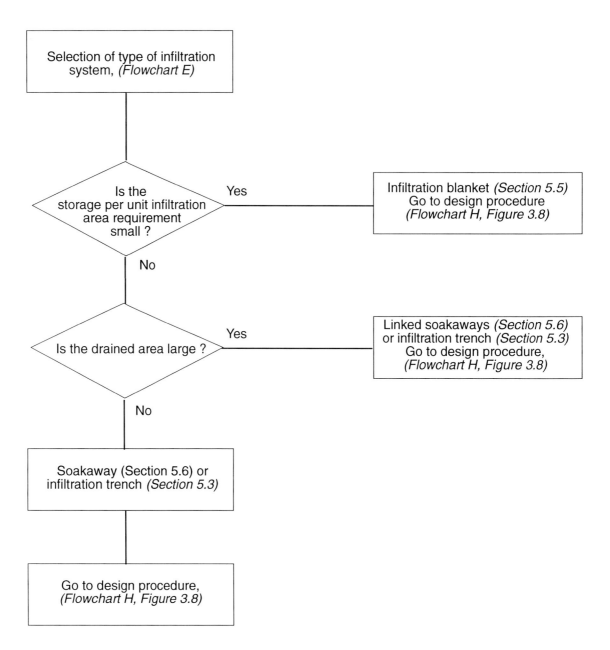

Figure 3.6 Flowchart F: Selection of type of sub-surface infiltration system

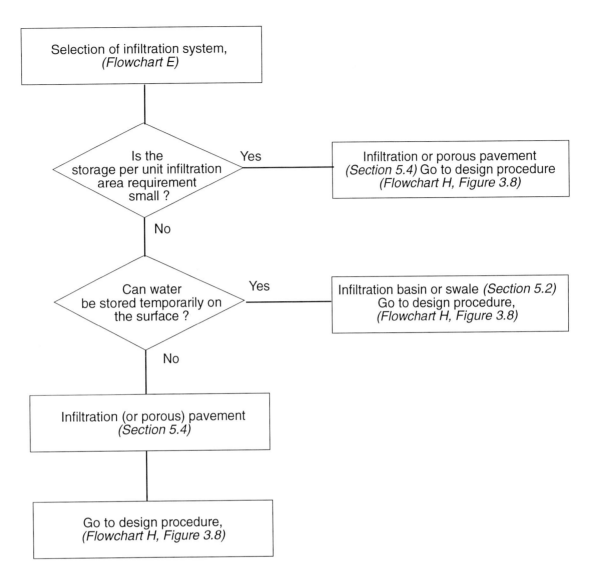

Figure 3.7 Flowchart G: Selection of type of ground infiltration system

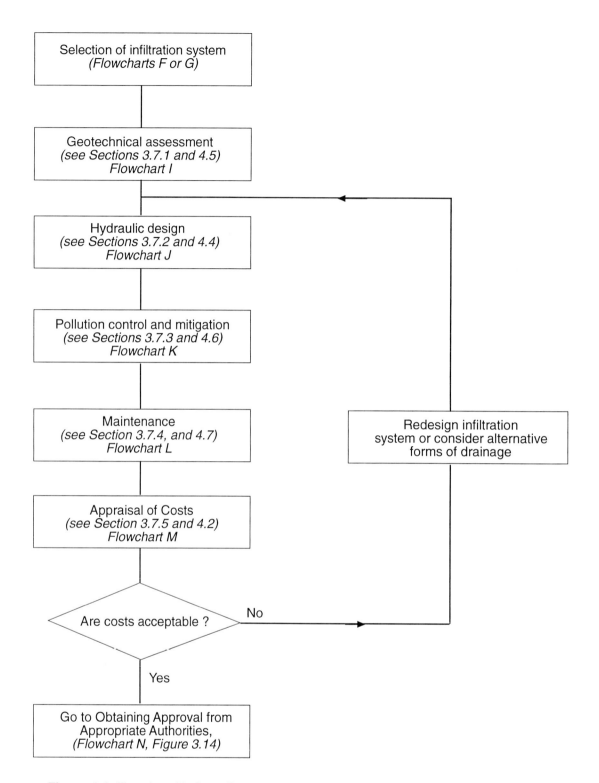

Figure 3.8 Flowchart H: Overall design procedure

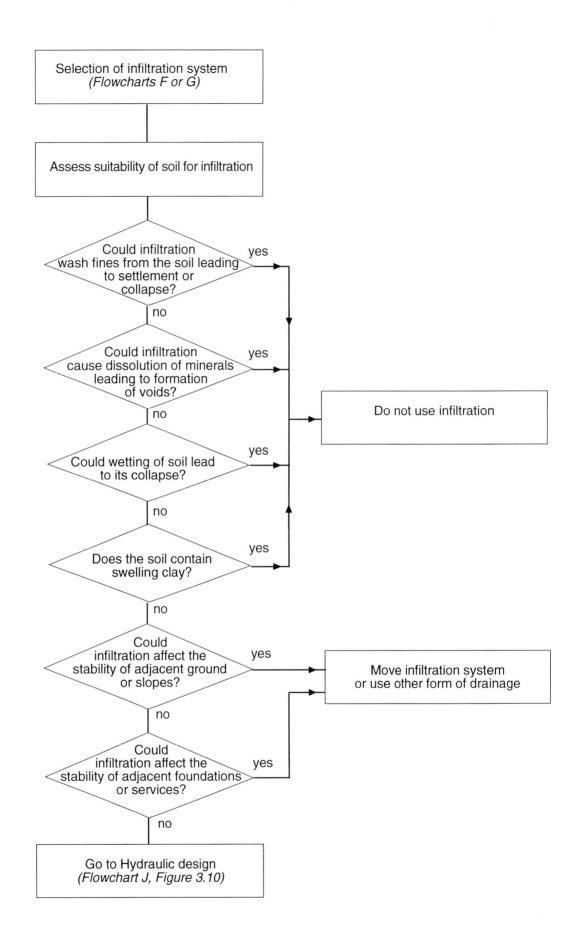

Figure 3.9 Flowchart I: Geotechnical assessment

3.7.1 Geotechnical Investigation

The process of geotechnical assessment is shown in Flowchart I, Figure 3.9.

Infiltration systems introduce water into the surrounding soil. In many cases this will have no significant effect on the soil and the system will work satisfactorily. In some cases, however, the introduction of water into the soil may have a significant impact on the ground around the infiltration system which can have serious implications for the stability of nearby services, foundations and slopes. The geotechnics should be appraised to assess the likelihood of problems. This should consider:

- an assessment of the nature of the soil and the impact of introducing water into it. The issues are whether this could result in settlement, erosion, chemical reactions or whether the soil is soluble.

- the location of the proposed system in relation to foundations, services and slopes and whether these may be affected by the presence of the infiltration system.

A further description of the geotechnical aspects of infiltration systems is given in Section 4.5.

3.7.2 Hydraulic design

The process of hydraulic design is shown in Flowchart J, Figure 3.10.

The purpose of the hydraulic design is to select the dimensions of the infiltration system which are sufficient to dispose of the runoff from storms of any duration with a selected return period. See Section 4.4 for details.

The method adopted for the hydraulic design of infiltration systems differs from those adopted in CIRIA (1992b) and BRE Digest 365 (1991). These are compared in Appendix 1.

3.7.3 Pollution control and mitigation

The process of pollution control and mitigation is shown in Flowchart K, Figure 3.11.

Water that enters an infiltration system may carry pollutants derived from the surfaces over which it has flowed. Consideration should be given to the nature of the pollutants, the quantity and their ultimate fate after they have entered the infiltration system.

The entry of sediment can lead to a reduction in the infiltration capacity of the soil. Wherever possible, therefore, sediment should be prevented from reaching the soil surrounding the system by the use of sediment traps or other means required by the NRA.

See Section 4.6 for more information.

3.7.4 Maintenance

The process of considering maintenance is shown in Flowchart L, Figure 3.12.

Any system will require some degree of maintenance throughout its life. Surface infiltration systems will normally require maintenance of the surface and sub-surface systems will also normally require clearing or cleaning. For some types of system this maintenance can virtually amount to reconstruction of the infiltration system. Effective maintenance can have a significant impact on extending the effective life of the infiltration system.

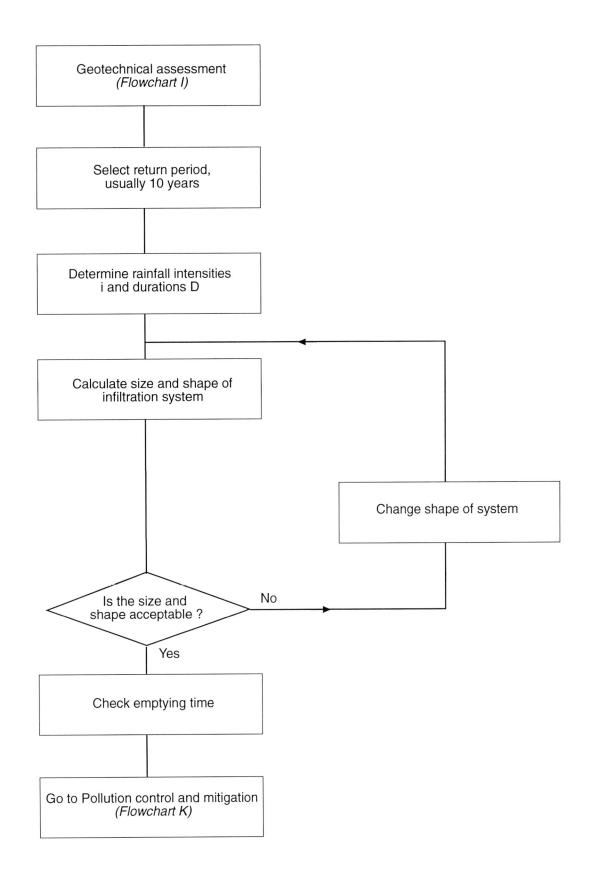

Figure 3.10 Flowchart J: Hydraulic design of infiltration systems

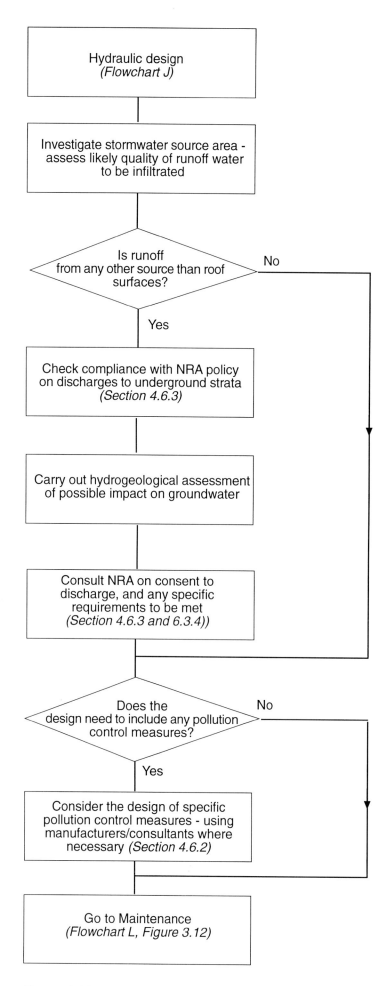

Figure 3.11 Flowchart K: Pollution control and mitigation

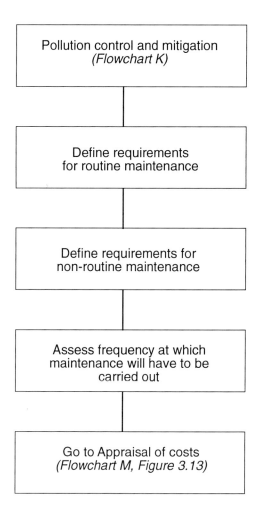

Figure 3.12 Flowchart L: Maintenance and adoption

Maintenance will usually be the responsibility of the owner of the land, unless other legal provisions have been made. It may be possible to arrange for the system to be maintained or adopted by an appropriate authority. If this is being contemplated the authority should be approached at the earliest possible opportunity. As how already been noted, at the time this manual was written it was the policy of the Statutory Sewerage Undertakers in England and Wales not to adopt infiltration systems.

For a general discussion of maintenance responsibilities see Section 4.7; for information on maintenance related to specific types of system the reader should refer to Section 5.

3.7.5 Appraisal of costs

The process of appraising costs is given in Flowchart M, Figure 3.13.

The designer will be concerned about the capital and recurrent costs associated with the proposed system. Where there are alternative methods of providing storm drainage, the designer will be particularly interested in making comparisons of the costs involved. In an appraisal to determine the overall most economical option, it is important that 'whole life' costs are considered. These include the operating, maintenance and replacement costs that will be incurred over a specified design life.

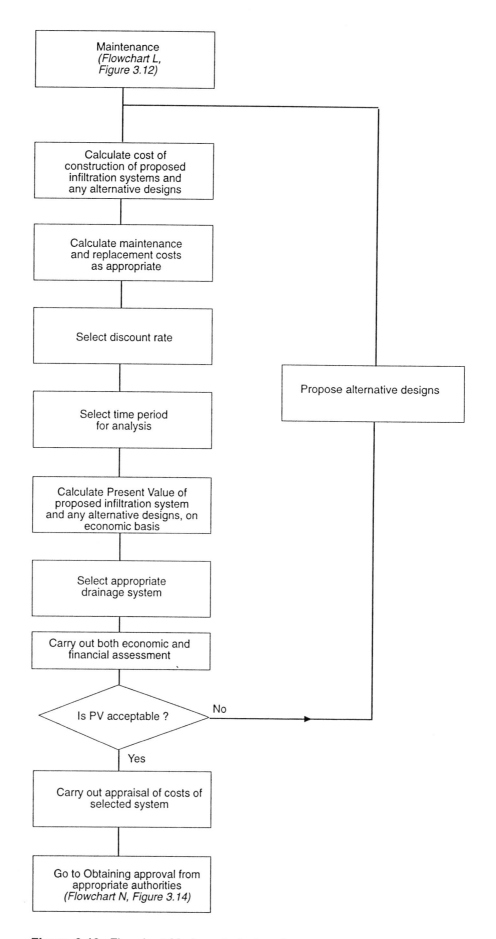

Figure 3.13 Flowchart M: Appraisal of costs

Section 4.2 presents a procedure for the determination of the whole life costs of a system to enable the cost of a proposed system to be assessed and comparisons to be made with alternative forms of drainage.

Where adoption is proposed, a financial appraisal should be carried out which will identify the costs appropriate to both the developer and the adopting authority. While the economic appraisal will identify the least cost option in terms of national resources, the financial appraisal will identify the cost implications for the various parties concerned.

3.8 APPROVAL FROM APPROPRIATE AUTHORITIES

The process of obtaining approval is given in Flowchart N, Figure 3.14.

Depending upon the nature of the proposed system, approvals will be required from various appropriate authorities. The procedures required are illustrated in Figure 3.14.

For a fuller description of these procedures, see Section 6.

3.8.1 Approval from the National Rivers Authority

If the discharge from the infiltration system is of a polluting nature a Consent to Discharge must be obtained. If it incorporates an overflow to a watercourse then, in addition, it will require Land Drainage Consent.

(a) Consent to Discharge

The NRA is the appropriate regulatory body to authorise a Consent to Discharge and will require to know at least:

- the site of the proposed infiltration system
- the nature of the discharge, and in particular its likely quality
- the expected volumes to be discharged.

It is advisable to discuss requirements with the NRA prior to making an application for Consent to Discharge.

Currently, the different regions of the NRA vary in their detailed practice but there is a progressive development of national guidelines which should result in greater consistency. There are some differences in organisation between the NRA regions, but as a general rule reference should be made to the Area Planning Liaison Officer.

(b) Land Drainage Consent

Where an overflow to a watercourse is proposed and thus a Land Drainage Consent is also required, the NRA will be the consenting authority in cases where the watercourse is designated 'main river'; otherwise the approval of the relevant District Council would normally be required. In the relatively few areas overseen by Internal Drainage Boards, its consent will be required if the discharge is to one of its supervised water courses.

3.8.2 Maintenance or adoption

If it is envisaged that in the future there may be a wish to pass the maintenance of the infiltration system to a sewerage undertaker or highway authority, it is advisable to consult them at the earliest possible stage to ascertain any requirements that they may have. This matter is discussed further in Section 4.7 on Maintenance Responsibilities.

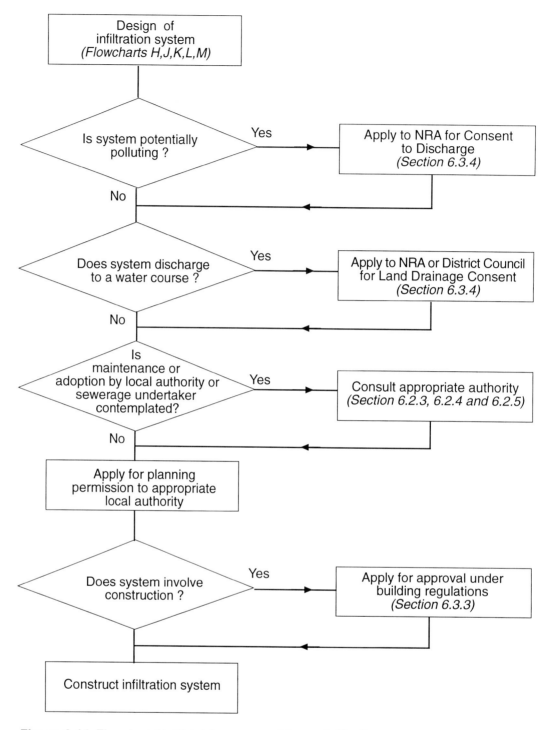

Figure 3.14 Flowchart N: Obtaining approval for an infiltration system

3.8.3 Planning permission

In general, planning permission will be required for an infiltration system. For a brief discussion of the legal aspects see Section 6. For a more detailed discussion the reader should refer to the report on legal aspects (CIRIA Project Report 25).

There is no direct guidance on how the planning authority should determine an application for planning permission. Past experience shows, however, that the authority will need evidence of the adequacy of the system to provide drainage and the likelihood of any flooding occurring.

It is likely that the fullest possible technical evidence will be required to support an application. It is recommended that this should include:

1. Size and nature of the area to be drained
2. Survey of ground conditions
3. Details of the infiltration system to be used
4. Results of infiltration test
5. Hydraulic calculations to demonstrate the adequacy of the system
6. Implications of storm events greater than the design rainfall.

If Consent to Discharge or Land Drainage Consent is required, the planning authority will normally have to be satisfied that these approvals have been obtained before they issue planning permission.

3.8.4 Building control

If the infiltration system requires any form of construction and if it is held to be part of the drainage system, it will need to satisfy building control. A building notice or full plans of the work will have to be submitted to the local authority or Approved Inspector. To obtain approval it is likely the fullest possible technical evidence will have to be presented, similar to that required for planning permission. Clarification of requirements should be sought from the local authority or Approved Inspector before formal application for consent.

4 General design considerations

4.1 INTRODUCTION

This section provides advice applicable to infiltration systems in general. It deals with appraisal of costs, site testing, hydraulic design, pollution control costs and maintenance responsibilities. Section 5 deals with specific aspects of different infiltration systems including construction. When considering the application of a particular type of infiltration system this section should be read in conjunction with the appropriate subsection from Section 5.

4.2 APPRAISAL OF COSTS

4.2.1 Introduction

This section is intended to provide guidance to engineers, developers and authorities on the appraisal of costs when comparing infiltration systems with other methods of drainage for stormwater. Further information will be found in CIRIA Project Report 24.

The section does not give detailed unit cost data because of the wide range of infiltration systems, each of which is site specific to the infiltration process and the site soils. Unit costs can be built up using manufacturers' prices and published unit rate data. The emphasis in the report is on guidance on how to determine the cost effectiveness of using an infiltration system in comparison with alternative forms of drainage system.

It is certainly true that infiltration systems can sometimes offer significant economies over positive drainage schemes. Care is necessary, however, in calculating the cost of an infiltration system and, if appropriate, in making the comparison with the cost of providing positive drainage. Strictly, the cost of conventional drainage of a new development and the enhanced system required downstream needs to be compared with the cost of infiltration drainage of the development and the (presumably) lower costs of any enhancements which may still be needed to the system downstream.

In considering costs, an economic or a financial view can be taken. An economic appraisal seeks to evaluate all the real costs and benefits to the community affected by a proposed development, while a financial appraisal is solely concerned with the costs, earnings and revenues directly applicable to a particular organisation.

Both economic and financial appraisals need to look not only at the initial capital costs but at the 'whole life' costs. These include the future operating, maintenance and replacement costs which will be incurred to enable the system to carry out its functions satisfactorily over a specified design life.

In an economic appraisal, the major difficulty is usually the assessment of the benefits and drawbacks of the scheme, for these may not be readily measurable in cash terms. This is especially so when considering environmental factors. For instance, the environmental benefits or drawbacks of encouraging surface water runoff to recharge the aquifers through infiltration systems require assessment of the possible pollution effect and the rise of the water table. These benefits or drawbacks can be complex to quantify in cash terms. If, for example, a developer proceeds against the warnings of the NRA of potential risks to groundwater quality, how will the developer's liability be incorporated into the exercise ?

In a financial appraisal the flows of money arising from the project are identified along with those who pay or receive these monies. This aspect of costs is most important when considering the options for a drainage system.

4.2.2 Perspective on costs

There will be important differences in the consideration of cost depending upon the perspective from which the appraisal of a drainage system is prepared. For instance:

- The *Site Owner/Developer* will, if the system is to be adopted, be interested in the minimum costs to meet the specifications for adoption of the system and the requisitioning costs of off-site works. He will only be interested in maintenance and replacement costs if he is not able to pass responsibility for operation and maintenance of the system onto an adopting authority. In order to convince an adopting authority, however, that an infiltration system is economical, the developer must expect to be able to demonstrate these costs.

- The *Authority or Sewerage Undertaker* which will operate and maintain the drainage system will be particularly interested in the operating, maintenance and replacement costs. In fact, the capital cost is not of particular interest to the adopting authority because this will be met by the developer. Their interest is in minimising future costs. It is likely that undertakers will favour the choice of a capital intensive system which has low future recurrent costs, despite the better overall economies of a system with higher maintenance and replacement costs.

Clearly, the adoption of a drainage system has a major impact on the analysis of costs and benefits and this becomes evident in a financial appraisal.

4.2.3 Costing of infiltration systems

(a) **Construction costs**

The costing of any piped system should be relatively straightforward. Pipes, manholes, gullies, and other related features can be costed from unit rates based on past prices or on prices obtained from published lists. Costing of concrete ring-type soakaways should, likewise, be straightforward once the appropriate location, construction and sizes have been identified. In a similar fashion the cost of construction of other infiltration systems, such as infiltration pavements or blankets, can be estimated by using standard costings for the individual elements of the system.

For surface infiltration systems there may be an additional cost related to the requirement for land which is dedicated to infiltration. If the infiltration system can be located on otherwise unused land for which the developer has title and access, and if the maintenance of this as open land is already a planning requirement, then logically in a financial appraisal there should be no cost allocation for using the land for infiltration. If it is necessary to purchase additional land for the infiltration system, this cost should be included.

(b) **Recurrent costs**

Replacement costs

When a drainage system fails to meet its design criteria, it is unlikely that the entire system will need to be replaced. It is more likely that certain elements can be rejuvenated or replaced, thus prolonging the design life of the system. For soakaway systems this may entail removing the structure and the surrounding granular material and soils which have become blinded by silt ingress and replacing with new granular material. For infiltration basins the silted up material on the bottom of the basin would be excavated and replaced with new granular material.

The frequency of replacement will depend on the estimated design life. An infiltration system may not have as long a useful life as a sewerage system and may therefore need to be replaced more frequently. For the purpose of a cost comparison between the two types of system, it will be necessary to allow for replacement costs during the design life of the entire drainage system.

Maintenance and repair costs

Maintenance and repair are carried out to ensure that the scheme continues to function efficiently as designed. A distinction has to be made between crisis maintenance and regular good practice maintenance. Many drainage systems are only maintained when problems arise — we are all aware of the gully pot which is only emptied when complaints are received that it is blocked.

It is known that soakaways are frequently ignored until failure occurs. However, those authorities who do undertake regular jetting of gullies and silt traps recognize that this will prolong the useful life of the soakaway. Annual cleaning of gullies and silt traps is therefore recommended, although in the design of the soakaway it would be safer to assume that cleaning is carried out only on a crisis basis leading to a shorter design life of the soakaway, especially those serving drainage systems containing gritty or soil laden water.

Infiltration basins or swales will require regular maintenance to control vegetation and undue siltation. However, this may well be covered by the landscaping costs of the area and, for the purpose of comparison of alternatives, maintenance may need to be apportioned arbitrarily.

Even a well-engineered and constructed positive surface water sewerage system will require some maintenance throughout its design life and regular inspections are desirable. Slow flowing, flat gradient sewers will be the most vulnerable to siltation problems. Again, it is common, (but not good practice) to organise maintenance on a crisis management basis and it has been assumed that sewer cleansing will only be carried out when severe backing-up or flooding occurs.

An indication of the cost of maintenance for use in cost appraisal is given Table 4.1.

Comparison with the cost of positive drainage

In many cases the immediate cost of an infiltration system is less than the alternative of providing positive drainage on the development site. In making such a comparison, however, it is important that the whole life cost is considered. This should take into account the costs of maintenance and, were appropriate, replacement, and also the effects of the drainage of the development on the existing system downstream.

The cost of providing positive drainage is normally sensitive to the distance to the nearest connection point of the existing drainage system and the adequacy of the existing system to take additional flows. Where the distance to the nearest suitable connection point is large it is likely that infiltration will be the cheaper option.

The companion report on costs describes an appraisal procedure which can be used to compare the costs of providing either an infiltration system or positive drainage. For the examples considered it is demonstrated that the cost of infiltration systems can be less than providing positive drainage, but it should be emphasised that the examples contain no allowance for costs arising from environmental concerns.

Table 4.1 Indicative costs of maintenance of drainage systems

System	Assumed Maintenance (a)	Indicative Cost (b) (Price base 1996)
Private drains and connections	Once yearly cleaning of gully pots and five yearly rodding of drains to clear blockages	Cost borne by private landowners
Local sewerage systems	Once yearly cleaning of gully pots	£2/gully/yr
	Jetting system, say once in 10 years	£1 per metre run of sewer
Sedimentation tanks	Once yearly jetting and cleaning	£50/sedimentation tank
Main and trunk sewers	Yearly visual inspection of manholes	£15/km
	CCTV survey of condition once in 60-year design life	£0.60/m
	Jetting once every 20 years	£1 per metre run of sewer

(a) This is the assumed maintenance and should *not* be regarded as recommended good practice.

(b) Figures are indicative only and apply to general maintenance of systems rather than dealing with incidents such as sewer blockages or soakaway failures where additional costs would be involved.

4.2.4 Cost appraisal

(a) Objective of procedure

The objective of the costing procedure is to obtain a 'like for like' comparison of the drainage scheme options in economic or financial terms.

In selecting a scheme, however, the indirect and intangible costs and benefits, although difficult to quantify in cash terms, should not be overlooked. For instance, an infiltration scheme may avoid disruption due to construction of sewers in built-up areas downstream of the site or may ameliorate water table deficiency in the aquifer region.

Where benefits or drawbacks are difficult to quantify in cash terms, they can be assessed by reasoned argument or by a points system related to defined objectives.

(b) Comparison methodology

As the purpose of cost appraisal is to evaluate the costs of alternative schemes, a cost-effective analysis is sufficient, each scheme being designed to provide as far as possible an equal standard of service.

In order to compare costs (capital and recurrent) it is therefore necessary to discount the costs from the time at which they are incurred to a common point in time. Hence the Present Value (PV) is determined for all costs over a suitable discount period.

The choice of period over which to carry out the discounted cash flow analysis is a matter of judgement. Periods of 25-30 years are commonly used for analysis, but the shorter the period, the more significant are the residual values of the components at the end of the discount period. With the use of spreadsheets on personal computers it is often more convenient to adopt a longer discount period than to evaluate residual values. To simplify analyses, the replacement periods should be whole fractions of the periods of analysis.

The design life of a positive system of drainage (say 60 years) is a convenient period over which to discount, even if not meaningful economically. For the purposes of comparison, it could be assumed that at the end of this period most of the components need replacing. The residual value of land may be the only significant value to affect the discounted costs.

Care should be exercised in assessing the residual value of sewers and manholes. Many systems will be able to continue service after the allocated design life but it should be appreciated that the cost of maintenance will increase and in many cases, where disruption to services, transport and reinstatement of carriageway is extensive, major costs could be incurred which may easily exceed the initial construction costs.

(c) Choice of discount rate

In order to determine the present values for comparison of scheme options, the discount rate must reflect the opportunity cost of the capital under consideration.

The discount rate for use in public service appraisal is known as the Test Discount Rate and is set by HM Treasury (1984). Private investors, however, may wish to set a rate more in line with the rate of return which they require in assessing the overall facility of a particular development project and this is likely to be at a higher rate. Potential adopting authorities will have their own current discount rates which they may require to be used if adoption is the objective. It should be understood that adoption of a high discount rate will favour options with low capital cost, short life and high recurring costs, whilst a low discount rate will have the opposite effect.

In looking at the most financially attractive scheme from a developer's perspective, it may be appropriate to use the discount rate that the particular firm uses to determine the opportunity costs of its capital and the guidance of the company's accountants should be sought.

It is advisable to repeat the analysis for discount rates greater and lower than the adopted rate to determine the sensitivity of the options to a change in discount rate.

4.2.5 Example of cost appraisal

Example: Housing scheme in East Anglia

Description of scheme

A comparison of costs between a positive drainage system and a soakaway drainage system was made for an eight-hectare development site. The site was located some 200 metres from the sea beyond cliffs which rise 20 to 30 metres above the beach. Despite being close to cliffs the site was perfectly suitable for soakaway drainage, having pervious strata which dipped away from the cliffs and posed no threat to cliff stability. No public sewer or watercourse was available within an economic distance of the site.

A comparison of the costs of a soakaway system of drainage and a positive system of drainage was carried out.

The soakaway system consisted of a system of collector sewers leading to 20 soakaways designed to handle the 1 in 10 year storm. Each soakaway drained, on average, an area of 350 m^2. Link pipes were provided to allow overflow from soakaway to soakaway in the event of an individual soakaway being unable to cope with excessive inflow and so providing a further safety factor.

The positive drainage system consisted of:

- an on-site sewerage network
- 400 metres of gravity sewer, 400 mm diameter

- a vertical drop shaft 25 metres deep and horizontal heading 30 metres long with concrete segmental lining and a ductile iron sewer pipe
- a ductile iron outfall pipe to the low tide mark.

Capital costs of schemes

Table 4.2 shows the capital costs of the alternative drainage solutions. The costs compared are those in which there were differences between the two schemes. 'On plot' costs of housing connections are not included, for instance, as these are common costs.

Table 4.2 Example – Comparison of capital costs of positive and infiltration drainage systems

Item	Costs £	
	Positive System	Soakaway System
ON-SITE CAPITAL COSTS		
Highway Gullies & Connections	11 680	15 013
Pipework on-site	31 249	10 509
Manholes on-site	16 292	1 250
Soakaways		21 735
TOTAL ON-SITE COST	**£59 221**	**£48 507**
OFF-SITE CAPITAL COSTS		
Sewers	25 924	
Manholes	3 520	
Reinstatement of Roads	8 034	
Outfall and Shaft	97 000	
Land Costs External to Site	20 000	
TOTAL OFF-SITE COST	**£154 478**	
TOTAL CONSTRUCTION COST	**£213 699**	**£48 507**
PROFESSIONAL FEES	26 712	6 063
ADOPTION FEES	5 342	1 212
TOTAL CAPITAL COST	**£245 753**	**£55 782**

Although the positive drainage option was unusual in having a large capital cost involved in the construction of the drop shaft, it can be seen that the soakaway system is still considerably less costly in terms of comparing on-site costs only. However, to provide a clearer indication of 'whole life costs' maintenance and replacement costs should be taken into account.

Replacement costs

The soakaways are unlikely to have a design life comparable with that of a sewerage system. The soils on this site were sands and gravels. Gully pots were provided on the estate roads and petrol/oil interceptors in the commercial areas. Blinding of the soakaway soils would be expected to be a reasonably slow process. A 30-year life to replacement was therefore a reasonable assumption. This would mean replacement of the soakaways once during the system design life of 60 years.

Assuming that the entire soakaway is replaced, which may not always be necessary, the PV of the replacement cost (£21 735) is therefore £5020, assuming a discount rate of 5% over a 30-year period.

Maintenance costs

In order to achieve a 30-year design life before replacement, the petrol/oil separator (acting as a silt trap) should be regularly cleaned and flushed out. This is costed as £50 per sedimentation tank per year for 20 soakaways, i.e. £1000 per year. Over a 30-year period this gives a PV of £18 929.

As good self-cleansing gradients could be achieved in gravity sewer design, maintenance costs of the positive drainage system are assumed to be minimal. However, it was considered that the vertical drop shaft and sea outfall would require at least an annual inspection and some routine maintenance to check pipework, etc, so an annual sum of £100 is allowed. Over a 30-year period this gives a PV of £1900.

Maintenance costs of gully emptying are assumed to be equal for both positive and soakaway systems and are therefore not included in the costing.

Whole life costs

Table 4.3 shows the present value costs of several scenarios, discounted at 5%.

A positive drainage scheme with an outfall at the site boundary, incurring no requisition costs would be the most economic. However, no account is taken of the effect of downstream peak flow impact. If these peak flows have to be channelled through an off-site sewerage system, such as 400 metres of sewer in the example, the costs of the positive system are less economic. If it is necessary to carry out any major off-site works, such as the drop shaft, the site contained soakaway system looks very attractive.

Table 4.3 Example – Comparative present values of drainage system options

Scheme	Present Values – 5% Discount Rate, Period of analysis 60 years			
	Capital	Replacement	Maintenance	Total PV
Positive system with drop shaft	£228 447		£ 1 900	£230 339
Positive system without drop shaft	£124 770		£ 1 900	£126 662
Positive system on-site cost only	£ 55 058		£ 1 900	£ 56 950
Soakaway scheme	£ 51 869	£ 5 020	£18 929	£ 75 818

The present values of the capital costs have been calculated assuming a two year construction period and that equal cost is incurred in each year (this accounts for small differences in figures between Tables 4.2 and 4.3).

It should be noted that the Replacement, Maintenance and Total PV figures are dependent upon the discount rate that is used. The selection of a different discount rate would alter the calculated PV values.

Conclusions

This example shows the importance of a detailed costing exercise and the inclusion of replacement and maintenance costs is clearly illustrated. The site specific nature of the costs is also illustrated by the need, in this instance, for a drop shaft and sea outfall and an off-site sewer for the positive system option.

In this case the adopting authority was reluctant to accept a soakaway system of drainage because it considered it would involve excessive maintenance costs and would eventually fail and have to be replaced. The cost comparison does show that these maintenance costs will be higher and that replacement of the soakaways will be required over the design life of the system, but that overall the soakaway system is less costly.

In the mid 1980s when this scheme was being considered, a commuted sum could have been arranged to cover these commitments and a potential developer would have preferred to make this commuted sum available rather than expend the considerable cost of a positive drainage system. However, under the Water Industries Act 1991 Section 146 adopting authorities can no longer require commuted sums for future maintenance costs.

Hence, if no legal mechanism is available to allow such a commuted sum to be paid, the developer may be forced to construct a highly expensive positive drainage system to meet the adopting authority's requirement of minimum maintenance and replacement cost.

4.3 HYDROLOGICAL AND HYDROGEOLOGICAL PRINCIPLES OF INFILTRATION

4.3.1 Hydrological processes

It is important, when designing an infiltration drainage system, to consider the hydrological and hydrogeological principles involved and to ask the following:

- What is the natural situation?
- Where will the water go?
- What possible hazards are there?

Rainfall landing on a natural ground surface can be transported by three routes (Figure 4.1):

- evapotranspiration back to the atmosphere
- runoff over the surface
- infiltration into the ground.

The proportions of rainfall taking each of these routes depends on such conditions as:

- meteorological conditions
- rainfall intensity
- ground surface slope
- type of soil
- ground cover
- soil moisture content.

An infiltration scheme will increase the quantity of runoff that is infiltrated and reduce the quantity that is disposed of in other ways.

4.3.2 Enhanced infiltration for stormwater disposal

It is evident that any hydrological change at the upper end of a catchment will have some effect at every point downstream. While the downstream effect of a single development may be small, the cumulative effect of many developments may be significant. Past practice has ignored this cumulative effect, with the result that flooding frequency in the downstream parts of a catchment has increased and dry weather flow in water courses has decreased. Greater attention has been paid in recent years to the cumulative effects of urban development, and the maximum outflow from new development has been frequently restricted to the equivalent outflow from the undeveloped site.

When an area is developed, impermeable surfaces such as roofs, roads, and pavements alter the proportions of evapotranspiration, runoff and infiltration of rainfall. Source control is used to try to restore this balance by either slowing down and storing the water in balancing chambers or detention ponds or by enhanced infiltration of the water into the ground using an appropriate infiltration drainage system (Figure 4.1). Many designs of infiltration drainage system are feasible, from large open vegetated areas to small discrete soakaways.

In most circumstances, the area of the infiltration system will be considerably smaller than the impermeable area being drained. Except for the most permeable of soils, the inflow rate to the infiltration system (product of the rainfall intensity and drained area) will exceed the outflow rate (product of the infiltration coefficient of the soil and the infiltration area). It is therefore necessary to store the water on-site and to allow time for it to soak away slowly through infiltration. Provision of sufficient storage capacity is essential for an infiltration system to perform properly. If the infiltration system is incorrectly designed, the outflow rate may not be enough to allow the system to empty sufficiently before the next rainfall event. The infiltration system will then overflow and the scheme will be deemed a failure. The design guidelines contained in this manual are intended to enable infiltration systems to be designed to acceptable factors of safety.

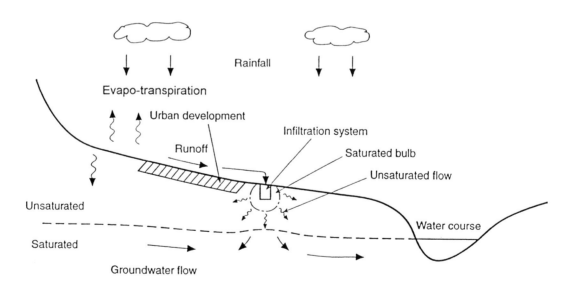

Figure 4.1 Infiltration system used for stormwater runoff control and disposal

4.3.3 Infiltration capacity of soils

For a soil to be suitable for accepting enhanced infiltration it must, in particular, be (a) permeable and (b) unsaturated. In addition, it must be of sufficient thickness and extent to disperse the water effectively.

(a) Permeability of soil

The capacity of a soil to infiltrate water can be described by using an infiltration coefficient. This is the discharge infiltrating into the soil divided by the area of infiltration. The infiltration coefficient of the soil is related to its permeability. This will be high for coarse grained soils such as sands and gravels, and low for fine soils such as silts and clays (see Table 4.4).

Table 4.4 Typical infiltration coefficients based on soil texture (Watkins, 1995)

Soil type	Infiltration coefficient (mm/h)
gravel	10 – 1000
sand	0.1 – 100
loamy sand	0.01 – 1
sandy loam	0.05 – 0.5
loam	0.001 – 0.1
silt loam	0.0005 – 0.05
chalk	0.001 – 100
cut off point for most infiltration drainage systems	0.001
sandy clay loam	0.001 – 0.01
silty clay loam	0.00005 – 0.005
clay	< 0.0001
till	0.00001 – 0.01
rock	0.00001 – 0.1

These figures may provide a useful first indicator of the magnitude of the infiltration capacity but the high ranges reported illustrate the importance of factors such as soil packing, soil structure, swelling clay content and the presence of fissures in rock which will significantly affect the infiltration capacity. It is necessary, therefore, to conduct a percolation test on-site to demonstrate the ability of the ground to accept infiltration in situ (see Section 4.3.5).

(b) Saturated soil

It is possible to dispose of water at a point beneath the water table but this practice, usually known as artificial recharge, requires special considerations which are outside the scope of this manual. Infiltration systems require an unsaturated soil to disperse the stormwater effectively and remove contaminants from it.

If the infiltration system is founded too close to the water table, a rise in water table during wet conditions could cause groundwater to enter the infiltration system, reducing the available storage (see also Section 4.4.2 (f)).

4.3.4 So where does the water go?

Water entering an infiltration system is temporarily stored. Eventually it soaks through the infiltration surface and percolates through the soil. Around a working infiltration system, a 'bulb' of saturation develops and the water flows through the soil under the influence of the hydraulic pressure gradient. As water seeps away from the infiltration surface, the flow area expands outwards and saturated conditions can no longer be maintained. The water continues to percolate through the soil as unsaturated flow, driven by capillarity and gravity.

Once the infiltration system is empty, the bulb of saturation will dissipate and the soil moisture will return toward ambient conditions.

The infiltrated water will continue to flow through the unsaturated soil, moving generally outward and downward. The water may by transported from the infiltration system via a preferential flow pathway due to geological conditions such as glacial outwash deposits, gravel infilled channels, solution cavities or rock fractures. Preferential pathways may cause problems, for example, if the water is led toward a building foundation.

With the issues previously considered in mind, the next step is to determine the infiltration capacity of the soil to estimate the size of system required.

4.3.5 Site determination of infiltration coefficient

(a) Performance of site test

The hydraulic properties of the ground are site specific. Some soils, such as sands, have high infiltration coefficients while others, such as clays, have low infiltration coefficients. The infiltration coefficient for a soil is an important element of the hydraulic design of an infiltration system. At present the only reliable method of determining the infiltration coefficient for a particular site is to carry out an infiltration test on site. The site test will provide an estimate of the infiltration coefficient but for the purposes of design there must be uncertainties associated with it. These uncertainties include:

- is the test pit representative of the full-size system?
- would the infiltration coefficient be different for different soil moisture conditions?
- will the infiltration coefficient reduce in time due to clogging of the system by fine sediments?

These uncertainties are allowed for in the design process by the incorporation of a factor of safety (see Section 4.4.2 (b)).

(b) Selection of site for test pit

The test pit should be dug at the proposed site of the infiltration system. If preliminary calculations indicate that a dimension of the infiltration system will be larger than 25 m a second test pit should be used. For larger systems further test pits would be required every 25 m. If the soil is fissured or there is reason to suspect that it may vary across the site, the distance between test pits should be reduced to 10 m. If more than one test pit is used, the mean value of the infiltration coefficients determined should be used in the design calculations.

(c) Size of test pit

The test pit should be at the same approximate depth as that anticipated in the full-size infiltration system. This implies that for a plane infiltration system, such as an infiltration pavement, a relatively shallow test pit will normally be required while for a large soakaway a deeper test pit is likely to be more appropriate.

The size of the test pit should be related to the size of the area to be drained. If the area to be drained is less than 100 m^2 the volume of water used in the test should be at least 0.5 m^3 and if the area to be drained is greater than 100 m^2 the volume of water used should be at least 1 m^3.

(d) Procedure for test

The test procedure described below is based on that described in BRE Digest 365.

The following is the procedure for carrying out a test:
1. Excavate a trial pit of the appropriate size.
2. Record the wetted area of the internal surface of the pit including both the sides and base when the pit would be half full of water.

3. Fill the pit with water to the invert level of the lowest incoming pipe.
4. At frequent intervals record the water level and time as the pit empties of water.
5. Repeat the test twice more in succession, preferably on the same day.

It is recommended that whenever possible the depth of water in the test pit should be comparable to the depth of water that is likely to occur in the completed infiltration system. This may not be practical, however, if the trial pit is deep and it would be difficult to supply sufficient water for a full-depth soakage test. Testing to the full depth of the trial pit may not be appropriate if, in the completed infiltration system, infiltration will only take place from the lower layers. Thus, for example, if an infiltration pavement is being considered then the results from a deep trial pit filled to its complete depth may not accurately reflect the infiltration from the completed system.

In these cases the test may be conducted at less than full depth. The calculation of the soil infiltration coefficient should then be based upon the actual maximum water depth achieved.

If necessary for stability the pit should be filled with granular material and a full-height perforated, vertical observation tube should be positioned in the pit so that water levels can be monitored.

(e) Analysis of test results

The time taken for the pit to empty from 75% to 25% of the depth of the pit should be determined, t_{p75-25}. The storage volume of the pit between 75% and 25% of the depth should be determined, V_{p75-25}. The area of the base and sides of the pit at 50% of the depth should be determined, a_{p50}.

The soil infiltration coefficient q is given by

$$q = \frac{V_{p75-25}}{a_{p50} \times t_{p75-25}} \tag{1}$$

Following the performance of a number of tests in the same size test pit the smallest value of q obtained should be used. The values of q may not necessarily reflect the indicative values given in Table 4.4.

4.4 HYDRAULIC DESIGN

4.4.1 Determination of design rainfall events

For the purpose of design calculations, it is assumed that the design rainfall hyetographs have a 'block' nature (constant intensity or flow during the duration of the storm) and that there is no attenuation of flow between the rainfall landing on the impermeable surface and the inflow to the infiltration system (Figure 4.2). The design rainstorm events can then be described in terms of intensity, duration and frequency and statistical values of these for a given location can be obtained from the Flood Studies Report (Institute of Hydrology, 1975).

For a given return period, the volume of runoff to the infiltration system is given by:

Volume of runoff = $Q.D = i.A_d.D$,

where:

i = rainstorm intensity (m/h)
D = rainstorm duration (h)
A_d = impermeable area drained (m²)
Q = inflow (m³/h)

Drainage systems are normally designed to a specific return period. The selection of the design return period will depend upon the consequences of failure and a return period appropriate to the risk should be selected. In many cases a return period of 10 years is used as a basis for design. This provides a similar level of protection as that provided by positive sewerage. A larger return period may need to be selected depending upon the potential consequence of failure or as a requirement for adoption by a suitable authority. At a particular location, for a specified return period, the rainfall depth varies with the duration of the storm event. This relationship between depth and duration varies throughout the country and so attention must be paid to the geographic location of the system. The Meteorological Office provides a service which includes the provision of rainfall statistics. Alternatively, the appropriate statistics may be calculated using published procedures.

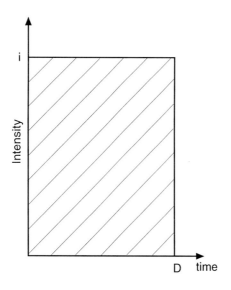
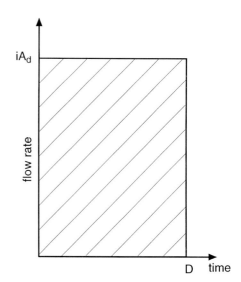

Figure 4.2(a) Rainfall hyetograph   **(b)** Hydrograph of inflow into infiltration system

The Institute of Hydrology has carried out an extensive analysis on rainfall statistics and has provided a method to determine the relationship between depth, duration and return period (Institute of Hydrology, 1975). This has formed the basis for the method described below.

The notation *MT-D* is used to identify a storm, where:

M is the depth of rain in mm
T is the return period in years
D is the storm duration

Thus *M*10-15 minutes is the depth of rainfall of a 10-year return period storm event of 15 minutes duration.

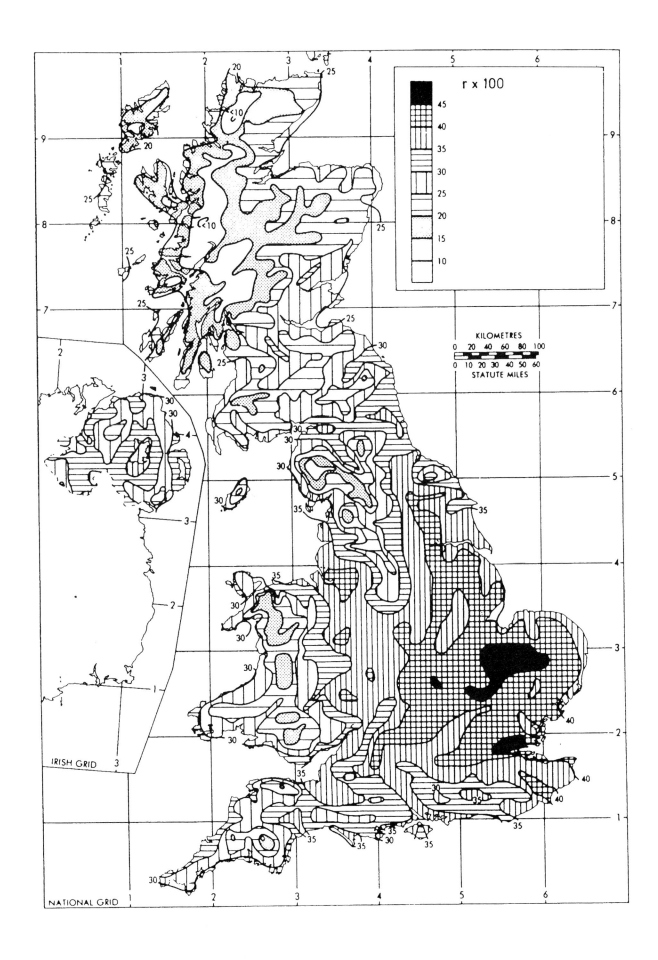

Figure 4.3 Values of rainfall ratio r in UK $\left(r = \dfrac{M5\text{-}60 \text{ min}}{M5\text{-}2 \text{ day}}\right)$

A design storm is assumed to be a rainfall event of duration D with a 10-year return period i.e. $M10$-D.

The average rainfall intensity, i, is obtained by dividing the rainfall depth by the duration.

Values of design rainfall depth, intensity and duration can be determined using Figure 4.3 and Table 4.5 for different storms with a 10-year return period as follows:

1. From the map in Figure 4.3 determine the rainfall ratio, r, for the location of the infiltration system, interpolating between contours. Note that r is the ratio of the 60 minutes to two day rainfalls of *five-year* return period.

2. Using the value of r from step 1 determine the *10-year* rainfall intensity, $M10$, for the required duration of storm D from Table 4.5, interpolating between the values if necessary. Where the other dimensions are in metres the rainfall intensity should be expressed in terms of m/h for consistency.

Repeat steps 1 and 2 for a variety of durations to obtain a set of values of i and D for 10-year return period rainfall events (see example).

4.4.2 Hydraulic design of infiltration system

(a) General

The primary purpose of an infiltration system is to dispose effectively of stormwater to the ground. The hydraulic behaviour of the system is therefore a dominant feature of its design.

There are no national standards relating to the design of infiltration systems, but BS 8301, BRE Digest 365, CIRIA Report 124 (Volume 3) and some Property Services Agency technical publications do cover the design of soakaways. Unfortunately, all of these documents differ in their design approach (see Appendix for some comparisons).One significant difference in the design approach adopted in this CIRIA Manual for infiltration systems is that it is assumed that infiltration takes place through the base of the system as well as through the sides.

The flow from an infiltration system is complex, but extensive modelling work at HR Wallingford has established that it is possible to make simplifying assumptions that produce results that are theoretically sufficiently accurate to act as a basis for design (Watkins, 1991). To apply these methods, infiltration systems are classified into two types, viz:

- **plane infiltration systems**, where the outflow from the system is predominantly through the base of the system, and

- **3-D systems**, where a significant amount of infiltration takes place through the sides of the system.

Examples of plane infiltration systems are permeable pavements and infiltration blankets, while examples of 3-D systems are soakaways and infiltration basins.

The procedure for carrying out a hydraulic design is shown in Figure 4.4.

(b) Factors of safety

In any design of an infiltration system it is not possible to have complete confidence in the specification of all the parameters. While some parameters, such as the area to be drained, will be known with some accuracy, others, such as the infiltration coefficient, will not be known with the same degree of confidence. In fact, one of the largest uncertainties is associated with the infiltration coefficient which may reduce over time, particularly if maintenance is poor. To account for this, a factor of safety is introduced into the design procedure which reduces the observed value of the infiltration coefficient. The factor of safety to be used depends upon the consequences of failure and engineering judgement is therefore required as to the factor to be used. The quoted factors of safety (see Table 4.6) take account of the possible loss of infiltration capacity throughout the design life of the system. Little information is available on such a loss of performance and so the figures given are suggestions rather than figures based on historic data.

Table 4.5 *M*10 rainfall intensity (mm/h) for duration D and ratio r

England & Wales

	Rainfall duration (D)									
	Minutes				Hours					
r	5	10	15	30	1	2	4	6	10	24
0.12	62.9	49.0	43.16	33.0	24.80	18.1	12.8	10.6	8.44	5.65
0.15	71.4	55.2	46.8	39.2	24.80	17.5	12.0	9.59	7.43	4.61
0.18	77.2	59.5	49.8	35.2	24.80	16.7	11.2	8.85	6.63	4.08
0.21	82.8	62.5	52.7	36.2	24.80	16.4	10.6	8.41	6.13	3.42
0.24	89.3	67.3	54.6	37.2	24.80	16.1	10.3	7.93	5.62	3.21
0.27	95.0	70.3	57.1	37.7	24.80	15.7	9.92	7.52	5.29	2.97
0.30	97.9	71.8	58.0	38.2	24.80	15.5	9.58	7.12	5.05	2.75
0.33	100.0	73.2	60.0	38.7	24.80	15.2	9.33	6.98	4.85	2.53
0.36	104.0	74.6	61.0	39.2	24.80	15.1	9.03	6.73	4.56	2.36
0.39	107.0	76.1	62.0	39.7	24.80	15.0	8.90	6.53	4.37	2.24
0.42	111.0	77.6	63.0	40.2	24.80	14.9	8.73	6.38	4.21	2.12
0.45	114.0	79.1	64.0	40.7	24.80	14.8	8.49	6.14	4.07	2.01

Scotland & N. Ireland

	Rainfall duration (D)									
	Minutes				Hours					
r	5	10	15	30	1	2	4	6	10	24
0.12	89.8	48.1	42.8	32.2	23.80	17.5	12.7	10.6	8.44	5.65
0.15	102.0	53.8	45.7	33.1	23.80	16.8	11.8	9.51	7.43	4.61
0.18	110.0	58.1	48.6	34.1	23.80	16.0	10.9	8.70	6.58	4.08
0.21	118.0	60.9	51.9	35.0	23.80	15.7	10.4	8.27	6.08	3.42
0.24	126.0	65.7	53.3	36.0	23.80	15.3	10.1	7.80	5.57	3.18
0.27	143.0	68.5	55.2	36.5	23.80	15.0	9.67	7.33	5.24	2.97
0.30	147.0	70.0	56.2	37.0	23.80	14.7	9.26	6.94	4.96	2.72
0.33	151.0	71.4	58.1	37.4	23.80	14.5	9.03	6.80	4.77	2.51
0.36	156.0	72.8	59.0	37.9	23.80	14.4	8.73	6.57	4.45	2.34
0.39	160.0	74.3	60.5	38.4	23.80	14.3	8.61	6.37	4.26	2.20
0.42	164.0	75.7	61.4	38.9	23.80	14.2	8.38	6.17	4.11	2.11
0.45	168.0	77.1	62.4	39.4	23.80	14.0	8.14	5.94	3.96	1.98

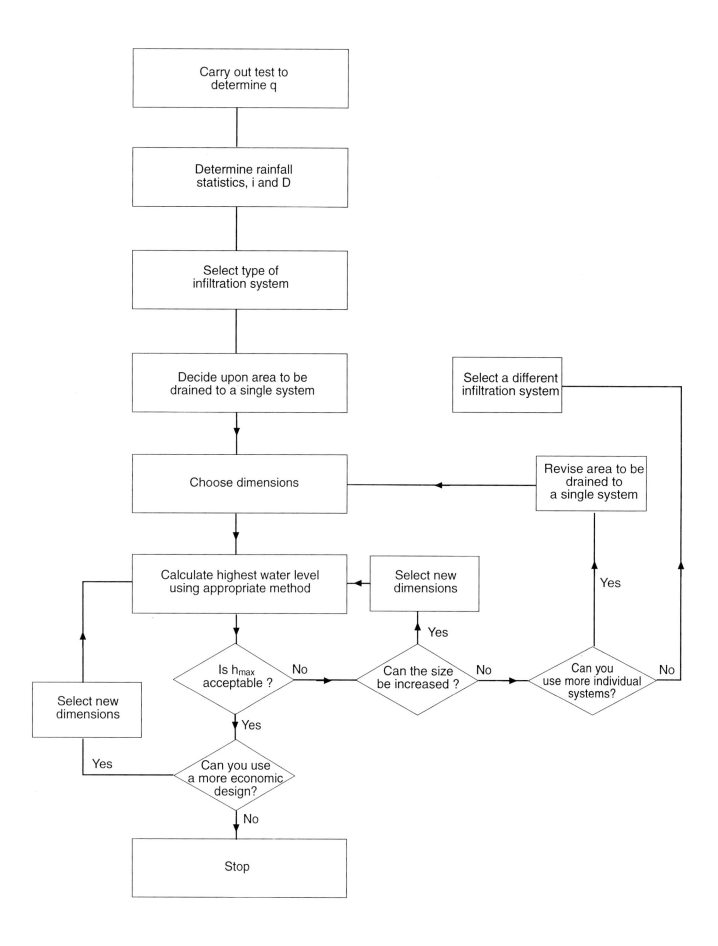

Figure 4.4 Detailed flow chart for hydraulic design

EXAMPLE

Determine the 10-year return period rainfall statistics for a site near Southampton

Step 1   From the map in Figure 4.3 the value of the rainfall ratio for the location near Southampton is $r = 0.35$.

Step 2   Using $r = 0.35$ determine the 10-year rainfall intensity for each storm duration from Table 4.5.

Storm duration, D (min)	Rainfall intensity, i (m/h)
10	0.074
15	0.061
30	0.039
60	0.025
120	0.015
240	0.009
360	0.007
600	0.005

The above method produces rainfall intensities of different durations with a 10-year return period. If rainfall intensities of a different return period are required growth curves described in the Flood Studies Report should be used. These relate the value for a 10-year return period to those for different return periods.

It should be noted that in the above the rainfall intensity is expressed in m/h. Rainfall intensities are also commonly expressed in mm/h, the ratio between the two values being 1000. In the subsequent hydraulic calculations, m/h has been selected as the units used should be consistent with each other. Thus if the size of the basin is expressed in metres and the infiltration coefficient in m/h then the rainfall intensity used in the calculations should be expressed in m/h.

Table 4.6 Factor of safety, F, for use in hydraulic design

Size of area to be drained	Consequences of failure		
	No damage or inconvenience	Minor inconvenience, e.g., surface water on car parking	Damage to buildings or structures or major inconvenience, e.g., flooding of roads
< 100 m²	1.5	2	10
100 m² to 1000 m²	1.5	3	10
> 1000 m²	1.5	5	10

The observed infiltration coefficient should be divided by the factor of safety to obtain the effective infiltration coefficient to be used in the hydraulic design.

In determining the infiltration area of the system, the area of the base of the system should be included together with the sides.

As a result of making a different set of assumptions from those used in BRE Digest 365 (i.e. no outflow through base of infiltration system) or CIRIA Report 124, it may well be that systems designed using the different methods may vary in size. Users should recognise that these variations are due to incorporating factors of safety into the design in different ways and do not reflect fundamental differences in the procedures used.

(c) Porosity of fill

If present, the porosity of fill or subgrade material needs to be taken into account. This may be obtained from laboratory tests or else the guide values provided in Table 4.7 may be used.

Table 4.7 Typical values for the porosity of fill material

Material	Porosity, n
Clean stone	0.4 – 0.5
Uniform gravel	0.3 – 0.4
Graded sand or gravel	0.2 – 0.3

A perforated concrete ring soakaway may be installed in a square or rectangular plan excavation and the annulus between the rings and the soil filled with clean stone. Under these circumstances an effective porosity, n', applies.

$$n' = \frac{\pi r'^2 + n\left(WL - \pi r'^2\right)}{WL} \qquad (2)$$

where r' is the radius of the ring sections
W is the width of the excavation
L is the length of the excavation.

If the clean stone filled section of the soakaway is sloping, then n' is a non-linear function of h and the soakaway must be treated in the same manner as for sloping sided systems.

(d) Plane infiltration systems

The required inputs are:

q infiltration coefficient, from percolation test (m/h)
A_D area to be drained (m^2)
n porosity of fill material (voids volume/total volume)
i,D intensity and duration of rainfall events with the required return period at the site location (m/h, h)
A_b base area of infiltration system (m^2)

The units specified for the variables are only suggested ones; it is essential, however, that the units of q, A_b, A_D, i and D are consistent with each other. Note that values of q are often quoted in mm/h (see Table 4.4).

For a given rainfall event discharging to an infiltration system of a particular size, the hydraulic equations can be solved to give the maximum depth of water, h_{max}. The equation for h_{max} is given by:

$$h_{max} = \frac{D}{n}(Ri - q), \qquad (3)$$

where R is the ratio of the drained area to the infiltration area, $R = A_D/A_b$

This equation can be solved for h_{max} by using a calculator or a spreadsheet on a computer.

The given procedure will ensure that stormwater will be able to infiltrate through the lower surface of the system into the soil at the required rate. For systems such as infiltration pavements which have a surface made of porous macadam or concrete blocks, a separate issue is the rate at which water can percolate through the surface. In such systems failure may occur because of poor surface infiltration and not due to poor infiltration into the surrounding soil. A designer should seek information to confirm the infiltration capacity of the selected surface material.

Procedure

Procedure to determine the maximum depth of water for a plane infiltration system.

The maximum depth of water corresponding to a particular storm duration is given by

$$h_{max} = \frac{D}{n}(Ri - q) \qquad (4)$$

1. Obtain the infiltration coefficient, q, by dividing the infiltration rate found from field tests by the appropriate factor of safety.

2. Find the porosity of granular fill material or estimate a value from Table 4.7

3. (i) Decide on the area to be drained, A_D, and the infiltration surface area, A_b.

   (ii) Calculate the drainage ratio, R, where $R = \dfrac{A_D}{A_b}$.

4. (i) Select a storm duration, D.
   (ii) Determine the corresponding rainfall intensity, i.

5. (i) Check whether q exceeds Ri. If so then the rate of infiltration exceeds the potential rate of runoff, in which case $h_{max} = 0$.
   (ii) Otherwise, calculate the value of h_{max} from equation (4).

6. Repeat steps 4 and 5 for a range of rainfall durations, constructing a computer spreadsheet or table of results.

7. Select the largest value of h_{max}.

A worked example is given in Section 5.4.3.

For an infiltration pavement, $R = 1$, step 3 is omitted and the maximum depth of water is given by

$$h_{max} = \frac{D}{n}(i-q) \qquad (5)$$

For an infiltration pavement where no subgrade material is provided to allow short-term storage of water, i.e. open lattice blockwork, storage occurs on open ground above the infiltration surface. In this case $R = 1$, $n = 1$, steps 2 and 3 are omitted and the maximum depth of water is given by

$$h_{max} = D(i-q) \qquad (6)$$

Alternatively, for an infiltration blanket the maximum depth h_{max} may be fixed and the designer may wish to know the base area of the infiltration system that will be required to ensure that the depth of water does not exceed h_{max}.

Procedure

Procedure to determine the base area required for a given maximum depth.

The equation for the base area A_b is given by equation (7)

$$A_b = \frac{A_D \, i \, D}{nh_{max} + qD} \qquad (7)$$

1. Obtain the infiltration coefficient, q, by dividing the infiltration rate found from field tests by the appropriate factor of safety.

2. Find the porosity of granular fill material or estimate a value from Table 4.7.

3. (i) Decide on the area to be drained, A_D.
   (ii) Decide on the maximum allowable water level, h_{max}.

4. (i) Select a storm duration, D.
   (ii) Determine the corresponding rainfall intensity, i.

5. (i) Calculate $A_D.i.D$, $n.h_{max}$, and $q.D$.
   (ii) Calculate A_b from equation 7.

6. Repeat steps 4 and 5 for a range of durations constructing a computer spreadsheet or a table of results.

7. (i) Find the largest infiltration surface area required.
   (ii) If this area is unacceptably large then increase h_{max} or decrease A_D and repeat from step 3.

A worked example is given in Section 5.4.3.

(e) 3-D infiltration systems

The required inputs are:

- q infiltration coefficient, from percolation test (m/h)
- A_D area to be drained (m²)
- A_b area of base (m²)
- n porosity of fill material (voids volume/total volume)
- i,D intensity and duration of rainfall events with the required return period at the site location (m/h, h), the type and shape of the proposed infiltration system and initial estimates of the dimensions (m).

The units specified for the variables are only suggested ones; it is essential, however, that the units of q, A_D, i, D and the structure dimensions are consistent with each other.

For a given rainfall event discharging to an infiltration system of a particular size, the hydraulic equations can be solved to give the maximum depth of water, h_{max}. The approach used depends on whether the facility has vertical or sloping sides.

Vertical-sided structures

This procedure can be applied to soakaways and infiltration trenches. The maximum water depth h_{max} in the infiltration system is given by the following equations:

$$h_{max} = a\left(e^{(-bD)} - 1\right), \qquad (8)$$

where

$$a = \frac{A_b}{P} - \frac{iA_D}{Pq}, \qquad (9)$$

$$b = \frac{Pq}{nA_b}, \qquad (10)$$

and P is the perimeter of the infiltration system.

These equations can be solved using a calculator, computer spreadsheet or graphically using Figure 4.5.

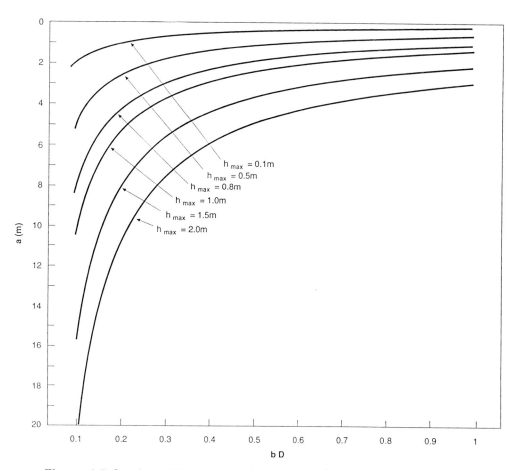

Figure 4.5 Graph to determine maximum depth for 3-D infiltration systems

Procedure

Procedure to determine the maximum depth of water in a given vertical-sided 3-D infiltration system.

Equation 8 is used to determine h_{max}

$$h_{max} = a\,(e^{(-bD)} - 1)$$

1. Obtain the infiltration coefficient, q, by dividing the infiltration rate found from field tests by the appropriate factor of safety.

2. Find the porosity of granular fill material or estimate a value from Table 4.7. If the structure is open, $n = 1$. If it is part-filled with gravel then the effective porosity, n', is used as provided by equation (2).

3. (i) Decide on the area to be drained, A_D.
   (ii) Choose the type and shape of infiltration system, i.e. cylindrical soakaway, infiltration trench.

4. (i) Select the proposed dimensions for the infiltration system, i.e. the radius of a cylindrical soakaway, the width and length of a rectangular plan system.
   (ii) Calculate the base area, A_b, and the perimeter, P, from the proposed dimensions.
   (iii) determine the value of b from $b = \dfrac{Pq}{A_b n}$.

5. (i) Select a storm duration, D.
   (ii) Determine the corresponding rainstorm intensity, i.

6. Determine the value of a from $a = \dfrac{A_b}{P} - \dfrac{A_D\,i}{Pq}$

7. Either calculate h_{max} from equation 8 or read off the value of h_{max} from Figure 4.5.

8. Repeat steps 5 to 7 for a range of rainstorm durations, using either a computer spreadsheet or a table of results.

9. (i) Find the largest value of h_{max}.
   (ii) If h_{max} is unacceptably high, return to step 4 and increase the dimensions.
   (iii) If h_{max} is still unacceptably high, either:

   (a) return to step 3(i) and reduce the area drained to an individual system
   or
   (b) return to step 3(ii) and choose a different type of system.

A worked example is given in Section 5.2.3.

Sloping-sided structures

For sloping-sided structures there is no simple analytical method for calculating the maximum water depth. A numerical procedure for calculating the depth is given in Bettess *et al* (1995). It is recommended that sloping-sided structures are approximated by a vertical-sided structure or that the method described by Bettess *et al* is used.

(f) Time for emptying

The hydraulic equations take both storage and infiltration into account and, if infiltration is small, will ensure that the system incorporates sufficient storage. If infiltration is too small, however, there is the possibility that the system will not have emptied following one rainfall event before the next rainfall starts. It has been suggested that, to ensure the system's readiness to deal with a rainfall event, the infiltration rate from the system should be sufficient to half-empty the storage in 24 hours. For a given geometry of system, this effectively imposes a minimum acceptable infiltration rate.

(1) Time for half-emptying a plane infiltration system.

The time for half-emptying a plane infiltration system is given by:

$$\frac{n \, h_{max}}{2q}$$

If the time for half-emptying is stipulated to be less than 24 hours and q is measured in m/h, then an acceptable infiltration coefficient is determined by:

$$q \geq \frac{n \, h_{max}}{48}$$

(2) Time for half-emptying a 3-D infiltration system.

The time for half-emptying a 3-D infiltration system is given by:

$$\frac{n \, A_b}{q \, R} \log_e \left[\frac{h_{max} + \frac{A_b}{P}}{\frac{h_{max}}{2} + \frac{A_b}{P}} \right]$$

If the time for half-emptying is stipulated to be less than 24 hours and q is measured in m/h, then an acceptable infiltration coefficient is given by:

$$q > \frac{n}{24} \, \frac{A_b}{P} \log_e \left(\frac{h_{max} + \frac{A_b}{P}}{\frac{h_{max}}{2} + \frac{A_b}{P}} \right)$$

(g) Proximity of the groundwater table

Water will continue to infiltrate out of an infiltration system even if the groundwater table approaches the base of the system or rises above the base. Thus infiltration systems will continue to operate even with high groundwater levels. There are, however, two disadvantages. First, if the groundwater rises above the base of the system storage is lost within the system. Second, the unsaturated zone surrounding an infiltration system is important in improving the quality of the storm runoff before it enters the main body of groundwater. Unless there are no concerns about possible quality of runoff, therefore, it is advisable to ensure that there is at least 1 m of unsaturated soil beneath an infiltration system.

The groundwater level may fluctuate throughout the year and this should be taken into account when selecting a level for the base of a system.

4.5 GEOTECHNICS

4.5.1 Introduction

Every infiltration device introduces water into the surrounding soil. In many cases this will have no significant effect on the soil and the system will work satisfactorily. In a few cases, however, the introduction of water into the soil may have a significant impact on the characteristics of the ground around the infiltration system and this can have serious implications for the stability of nearby services, foundations and slopes.

The most important factors that increase the potential for problems are:

- the volume of water that can be expected to infiltrate into the soil
- the rate of infiltration of water into the soil
- the soil type and its reaction to the addition to water
- the proximity of other engineered structures.

To avoid potential problems, it is recommended, therefore, that as part of any design of an infiltration system, a geotechnical appraisal should be carried out. This appraisal should firstly assess established local practice, in the light of the magnitude of the proposed infiltration system. The appraisal should also bear in mind the proximity of adjacent structures and hence the likely importance of the consequences of any problem that may arise. The appraisal should then consider the issues noted in Section 4.5.3 in the light of the available data.

4.5.2 Data collection

In considering the potential geotechnical problems at a site, the following information should be sought:

1. The geotechnical history of the site.
2. Information on groundwater levels.
3. The way in which the ground will react to water entry.
4. The location of adjacent foundations, slopes and services, including their robustness to movement.
5. Any obstructions to natural water movement in the ground caused by the local geology or nearby structures.

4.5.3 Geotechnical issues

Effective stresses

An increase in the pore water pressure in the soil may reduce the effective stresses, at constant total stress. This leads to a reduction in the shear strength of the soil which may in turn affect the stability of slopes and the bearing capacity of foundations, for example:

Loss of bearing capacity of adjacent foundations

As noted above, the entry of water into the ground from an infiltration system may reduce the load bearing capacity of adjacent foundations. This can lead to increased settlement of the structure and, at worst, complete failure. Infiltration systems should normally be built at least 6 m away from the nearest foundation of any buildings. Particular problems may arise with foundations in chalk, fill material or other soils liable to collapse, and under such circumstances the advice of a geotechnical expert should be sought as to the advisability of using an infiltration system.

Slope stability

The entry of water into the ground from an infiltration system may reduce the stability of a slope. Thus the use of an infiltration system close to a slope may cause the slope to fail.

If an infiltration system is to be constructed in the vicinity of a slope, it is essential that a geotechnical assessment is made of the impact of water entry on the stability of the slope.

Ground movement

Heave may occur as a result of the entry of water into the ground from an infiltration system. Such a phenomenon occurs in clays or clayey soils. Alternatively settlement may be caused which, given the non-uniform nature of natural ground, is not likely to be uniform and may adversely affect the system or the surrounding area. It may, therefore, be necessary to estimate the likely magnitude of differential settlement across the proposed infiltration system and accommodate such movement in the design.

Erosion of soils due to seepage of water

Depending on the pressure gradient and relative sizes of particles, seepage of water through soil may cause an erosion of the fine particles leading to the creation of voids within the soil matrix. This in turn may lead to settlement and collapse. When a seepage line emerges on a slope face, for example, the potential for water to wash away the fines is realised.

Dissolution of soil due to seepage of water

The entry of water from an infiltration system into a soluble material may lead to the dissolution of the minerals in the soil. This may produce voids in the soil leading to related problems. Calcareous soils or rocks are particularly prone to such problems and the use of infiltration systems in these materials is not advised.

Soils which chemically react with water in other ways

In most soils there is no significant impact on the chemical composition of the soil when water enters during infiltration. For a few types of soil, however, water entry into the ground can result in chemical transformations which have a significant impact, causing changes in stability and settlement characteristics. Water entry may cause anhydride to change to gypsum resulting in large expansion pressures and heave. Minerals such as pyrite also react with water.

Collapsible soils

Though the majority of soils in the UK are suitable for infiltration some soils, when they are wetted, undergo a radical re-arrangement of the soil particles. This can lead to a decrease in volume even if the load on the soil is unchanged. Examples of soil types that have this characteristic are: many loose fills and alluvial, residual, colluvial and gypsiferous deposits. In the UK the most common problem has arisen with loose fills where water entry can lead to spectacular collapse or settlement which may damage nearby structures. At the design stage the question should be asked whether the ground is well compacted and stable, or whether water entry will lead to the collapse settlement of the soil which could affect adjacent structures or their services.

The use of infiltration systems in loose fills or known collapsible soils is not recommended.

Highly swelling clays

Some of the soils in the UK are highly shrinkable/swelling clays. Typically such soils occur in the Southern and Eastern part of the country. They may be identified from local experience, for example, as giving rise to foundation subsidence at a time of drought, particularly in the vicinity of trees. The addition of water to such initially unsaturated soils, may cause large increases in volume, equally adversely affecting the foundations of dwellings. The use of infiltration systems in the vicinity of such foundations is not recommended.

Settlement due to mineral extraction

Ground may be stable now but future movements may be induced as a result of past or future mineral extraction. In the UK the most common cause of such movements is associated with coal extraction. Guidance on the magnitude of movement can be found in the *Subsidence Engineer's Handbook* (1975). Any infiltration system located in ground that may experience mining-related movement may suffer damage, particularly the connection of any drains which may be sheared and displaced. Any such settlements would also adversley, affect any alternative drainage system.

Construction

In situations where the installation of the infiltration system requires excavation, consideration must be given to the stability of the sides of the excavation particularly during its construction.

A significant hazard to the future effectiveness of infiltration systems can occur during construction or when there is construction in the immediate area (Hogland *et al*, 1987; Diniz, 1980). During that time an extensive portion of the land area may be unprotected from soil erosion. If this material is washed onto the infiltration surface rapid clogging may occur. There is also a danger that construction vehicles may compact the surface and reduce its permeability. To minimise these risks the following steps are recommended:

1. Locate infiltration systems away from roads.
2. Potential sources of clogging material, such as traffic, cement mixing and vehicle maintenance should be located away from areas to be used for infiltration.
3. Runoff which could be sediment laden should be kept away from the infiltration system during construction (Stahre and Urbonas, 1990).

If significant landscaping is to take place in the vicinity of the site of the infiltration system, the hydraulic design should be based on an infiltration test carried out after the landscaping has been completed.

Geotextiles

As noted above, seepage of water through soils may cause 'fines' to migrate, leading to the creation of voids. This can reduce stability and increase potential for settlement. Geotextiles are normally used in infiltration systems to separate different materials, and to limit the extent of this migration.

The full specification of the geotextile should take into account the purpose which it is intended to fulfil.

Geotextiles may be used to prevent penetration of a finer grained soil into a coarser material. A geotextile is required if the coarser material does not meet the filter requirements to prevent movement of the finer sediments into the coarser material. In such cases the replacement of the geotextile is either difficult or expensive. Flow must, therefore, pass through such a geotextile without clogging and satisfy appropriate retention and permeability criteria.

4.5.4 Checklist for design of infiltration systems

1. Are there any structural foundations close to the proposed site? If yes, ensure that the infiltration system is at least 6 m from the foundations. If the soil is a chalk, a fill or other soil liable to collapse, seek the advice of a geotechnical expert.

2. Is the infiltration system near a slope? If yes, then consider the impact of raised pore water pressures on the slope stability.

3. Could settlement affect the system or the surrounding area? If yes, estimate the likely magnitude of movement and accommodate in the design.

4. Will seepage of water through the soil cause erosion of the soil particles? If yes, either resite the infiltration system or seek to control the seepage route via drainage to avoid erosion.

5. Is the ground soluble? Will the entry of water cause the minerals in the soil to be dissolved and hence washed away? Is the ground a chalk soil or rock? If yes, the use of an infiltration system is not recommended.

6. Will the soils react chemically with the infiltrating water in other ways? If yes, the use of an infiltration system is not recommended.

7. Is the soil liable to collapse? Is it a loose fill material? If yes, the use of an infiltration system is not recommended.

8. Are there 'highly swelling' clays in the vicinity of the infiltration system? If yes, the use of a system in the vicinity of foundations is not recommended.

9. Could ground movements occur in the future that might affect the site? If yes, adopt a flexible design approach that will allow remedial action at a future date.

10. Can sediment entry into the system be reduced or prevented? If so, take the necessary preventative measures.

11. Can sediment entry into the soil be reduced by the use of geotextiles? If so, steps must be taken to allow their cleaning or replacement.

4.6 POLLUTION CONTROL REQUIREMENTS AND METHODS OF MITIGATION

Any water that enters an infiltration system will carry pollutants derived from the surfaces over which it has flowed. Consideration needs to be given to the nature of the pollutants, the quantity and their ultimate fate after they have entered the infiltration system. For a more detailed discussion of the sources and methods of control of pollution from new and existing highways the reader should consult CIRIA Report 142 (CIRIA 1994).

Further information on the water quality aspects of drainage systems can be found in:

* *Design of flood storage reservoirs*
  M.J. Hall, D.L. Hockin and J.B. Ellis
  CIRIA Book 14, 1993

* *Control of pollution from highway drainage discharges*
  CIRIA Report 142, 1994

* *Design and management of constructed wetlands for the treatment of wastewater*
  CIRIA Funders Report FR/CP/34, 1996

* *Use of industrial by-products in road construction: Water quality effects*
  CIRIA Funders Report FR/IP/11, 1996

4.6.1 Sources of pollutants

The type and concentration of pollutants in the water will depend upon the nature of the catchment surfaces over which the water has flowed. The quantity and character of pollutants resulting from runoff from roof areas, residential and public amenity land, highways and car parks, and industrial areas, can vary greatly.

The type and quantity of pollutants mobilised during a storm event is usually highly dependent upon the depth and intensity of the rainfall. In some cases, however, the pollution is primarily dependent upon other factors, such as regular maintenance work or accidents. Example situations in which runoff may be affected include:

- regular maintenance work, e.g. the application of herbicides on highways and public areas, or use of de-icing chemicals on roads and at airfields

- accidents, e.g. result of accidental spillages on roads and yards, or the ensuing cleanup operations.

There is some information available on the quality of runoff from different types of surface. Table 4.8 shows data on the quality from four roads (Urban 1,2,3 and Rural) compared to the quality of rain. All such data, however, have to be viewed in the context of the particular circumstances in which they were collected and care should be taken in applying them to other situations. As Table 4.8 shows, the variation even at a single site can be significant.

It is difficult to provide a satisfactory context for these figures. A natural form of comparison would be with common environmental standards. Unfortunately, the EC Groundwater Directive only lists substances and gives no quantitative standards. The EC Drinking Water Directive and the Dangerous Substances Directive do provide quantitative standards but a comparison with these is potentially misleading. The runoff analysed in Table 4.8 will inevitably be diluted to an unknown extent. Furthermore the figures in the table refer to concentrations during storm events whereas the standards in the Directives should be more closely related to annual averages.

4.6.2 Removal of pollutants

Pollutants can be removed from the runoff or their quantity reduced by physical and/or biological methods. These methods can be used before the water enters the infiltration system or as part of the system itself.

(a) Physical methods

The physical methods include:

- detention basins
- oil, petrol or grease interceptors
- sedimentation tanks or ponds
- infiltration pavements
- sand filter beds.

Studies have been carried out on the removal efficiencies of reed-beds and swales which operate partly on the basis of physical processes and partly biological. These indicate that such systems can be very effective in removing many types of pollutants. For more information see CIRIA Report 142 (CIRIA 1994) and Ellis (1991).

It should be emphasised that the performance of such systems is dependent upon the amount of maintenance that is carried out and that effective removal of pollutants is normally only achieved if satisfactory maintenance is performed.

Table 4.8 Reported ranges of pollutant levels in runoff from roads found in various locations (Colwill *et al* 1984; Strecker *et al* 1990).

Pollutant	Rainfall	Urban I	Urban II	Urban III	Rural
Electrical conductivity (μS/cm)	8-80	6-20 000	-	-	-
Total solids (mg/l)	18-24	145-21 640	11-40	-	-
Total dissolved solids (mg/l)	-	66-3050	-	-	-
Total volatile solids (mg/l)	-	12-1600	-	-	-
Volatile suspended solids (mg/l)	-	12-1500	-	20-78	6-25
Total suspended solids (mg/l)	2-13	2-11 300	-	68-295	12-135
Oil/hydrocarbons (mg/l)	-	0-400	3-31	-	-
COD (mg/l)	2.5-32	5-3100	-	57-227	28-85
Chloride (mg/l)	1-11	4-17 000	4-27	-	-
Bromide (mg/l)	-	0.02-6.0	-	-	-
Total lead (μg/l)	0.024-10.4	10-14 500	10-150	102-1562	24-272
Total zinc (μg/l)	0.02-4.9	1 000- 15 000	20-1900	192-564	35-185
Total cadmium (μg/l)	0.013-0.056	2-400	-	-	-
Total copper (μg/l)	0.06-0.48	7-2500	10-120	25-119	10-50
Total chromium (μg/l)	0.023-0.08	18-270	-	-	-
Total nickel (μg/l)	-	20-1500	-	-	-
Total organic carbon (mg/l)	1-18	5-120	-	8-74	3-17
Nitrate & nitrite (mg/l)	0.01-5.0	0.3-6.9	-	0.4-1.5	0.2-0.9
Total nitrogen (mg/l)	0.5-9.9	0.2-14	0.2-1.0	1.0-3.2	0.3-2.2
Total phosphorus (mg/l)	0.001-0.35	0.3-4.4	-	0.2-1.0	0.1-0.5
BOD (mg/l)	1-15	25-700	8-25	-	-

Figure 4.6 shows examples of simple oil and petrol interceptor designs while Figure 4.7 shows how they can be incorporated into an infiltration system. There will be a forthcoming British Standard for the design of a light-liquid separator produced under the auspices of BS Sub-committee B/505/8.

Work on sediment traps has indicated that they can be effective in retaining particulate pollution. This is particularly so for the large particle sizes. Trapping efficiencies for particles over 300 μm can be as large as 95% but this may fall to as low as 15% for particles smaller than 60 μm. Where a sediment trap permanently retains water then the quality of the retained water may deteriorate leading to the release of pollutants during the next rainfall event.

The efficiency of sedimentation tanks or detention basins depends upon the size of the tank or basin relative to the incoming flow. This can normally be usefully approximated by the residence time. Typical removal rates measured for detention basins in the USA are between 80 and 90% for total suspended solids, 70-80% for lead, 40-50% for zinc and 20-40% for BOD or COD.

Infiltration or porous pavements may also enhance water quality. The principal mechanisms of pollutant retention are thought to be sedimentation, filtration and chemical adsorption onto materials within the structure, (Pratt, 1989). This latter effect seems to be dependent upon the material used for the sub-base. Studies by Hogland *et al* (1987) indicate that infiltration pavements can be very effective in retaining particulate pollutants. There is no information available, however, on the impact on other forms of pollutants.

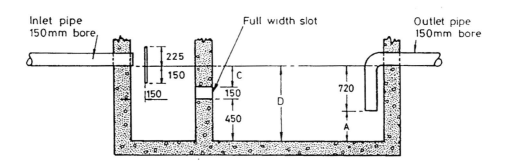

2-Compartment oil interceptor

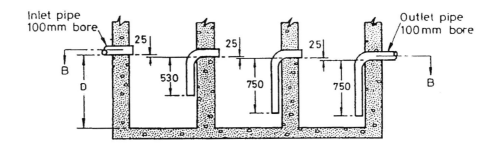

3-Compartment petrol interceptor

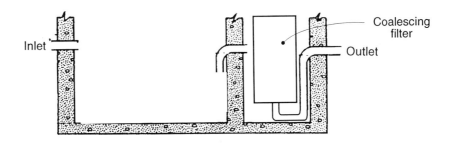

Coalescing interceptor

Figure 4.6 Examples of simple oil and petrol interceptor designs

Regular maintenance is necessary for such physical methods to continue to be effective in removing sediments and associated pollutants.

The traditional designs of oil or petrol interceptors are not effective at dealing with tertiary butyl methyl ether (TBME) which is a constituent of unleaded petrol. TBME, which is water-soluble, causes taste and odour problems in water in very low concentrations. To overcome this problem it is necessary for the 'first flush' of a storm to be prevented from entering the infiltration system by retaining it in storage. Arrangements need to be made to remove the oil and petrol from the interceptor before it is full.

(b) Biological methods

Biological methods may be used to remove or reduce the quantity of pollutants. These methods can be used before the water enters the infiltration system or as part of the system itself.

Biological methods include the use of:

- wetlands or ponds
- reed-bed systems
- grassed swales.

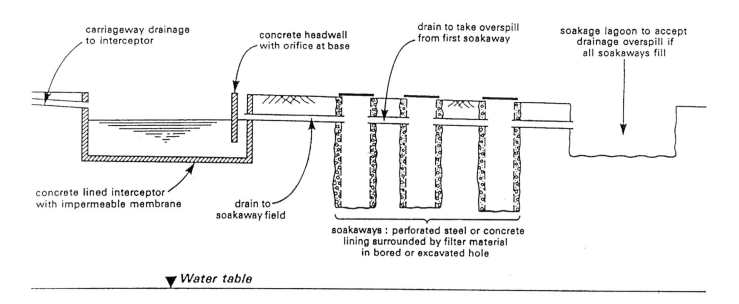

Figure 4.7 Example of a highway runoff infiltration system with an interceptor in place (after Price, 1994)

Biological methods can have a significant impact on the level of BOD, suspended solids and heavy metals. The efficiency of such methods depends upon the area provided, but may also vary throughout the year depending upon the growing season of the plants. To protect the biological environment, pre-treatment of stormwater will be necessary where significant quantities of oil are expected. Figure 4.8 shows a cross-section of a constructed wetland for water treatment. For more information see CIRIA Report 142 (CIRIA, 1994) and Ellis (1992).

Wetlands appear to control pollution through a number of mechanisms, including bio-filtration, sedimentation, adsorption, biological uptake and physio-chemical interactions. There are limited data available on the impact of reed-bed systems, on water quality for example, but the data indicate that significant improvements in water quality can be achieved. The performance of reed beds will depend upon the species present at the time of year and the flow rate; Ellis quotes typical figures of nearly 90% for removal of BOD for example. For more information on the effectiveness of such systems see Ellis (1991).

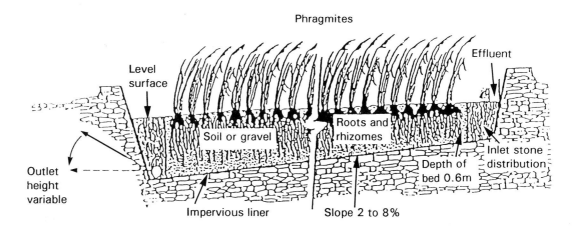

Figure 4.8 Typical arrangement for a reed-bed treatment system

Biological methods imply a long-term maintenance commitment to maintain their effectiveness. Periodic removal of sediment and harvesting of vegetation will be required. As the capacity of such systems to incorporate pollutants is finite, the useful life of the system depends upon a number of factors. If will, therefore, be necessary from time to time to replace the soil and vegetation in the basin or pond, although the life of the system can be extended by careful management.

These methods do not treat non-biodegradable pollutants, including many herbicides. Where infiltration methods are used for highway drainage a biodegradable herbicide such as Glyphosphate should be employed for weed control on highway verges. In view of the conflict between the use of herbicides and a duty of care for the environment, the wisdom of using herbicides in any but the most essential circumstances should be questioned by all highway authorities.

4.6.3 Groundwater protection

Depending upon the location of an infiltration system, there may be concern that pollution of an aquifer resource could occur. The important factors are the likely pollution loads in the stormwater, normally categorised in terms of the type of surface being drained, and the location of the infiltration system with regard to a water supply source or an aquifer. An infiltration system should normally comply with the NRA policy on discharges to underground strata.

The NRA Groundwater Protection Policy (GPP) (NRA, 1992) is based on the concept of 'groundwater vulnerability' which is defined simply as a measure of the ease with which unacceptable effects on groundwater resources can occur. The GPP describes aquifer protection in terms of both 'source' and 'resource' protection. The NRA will seek to reduce the risk of groundwater pollution by controlling activities likely to cause pollution, where vulnerability is greatest.

Vulnerability is dependent on the physical situation at a specific location and, therefore, the vulnerability of a resource to a particular pollutant threat will depend on: the nature of the solid and drift geology; the overlying soil characteristics; and the depth of the unsaturated zone. Although legislation requires that all groundwater be protected, regardless of use, the practical need to give priority to resources which are already being exploited for supply means that an additional factor needs to be considered where a groundwater resource is in use, i.e. the proximity of the point of infiltration to a source of abstraction such as boreholes, springs and wells. Hence the policy of 'Source protection'. This factor can be important, for instance, where degradable pollutants are involved. 'Source protection' is regarded as additional to the more general considerations of overall 'resource protection'.

Table 4.9 NRA Acceptability Matrix for Discharges to Underground Strata (from NRA Groundwater Protection Policy, 1992)

	SOURCE PROTECTION		
Activity	I Inner Zone*	II Outer Zone*	III Catchment Zone*
C. DISCHARGES OF SURFACE WATER TO SOAKAWAY FROM:			
Roof Drainage	No objection (R5) (provided for sole use of roof drainage)	No objection (R5)	No objection (R5)
Impermeable Areas			
- public/amenity	Not acceptable (R1)	Acceptable (R4)	Acceptable (R4)
- large car parks	Not acceptable (R1)	Acceptable (R3/4) (with interceptor)	Acceptable (R4) (with interceptor)
- lorry parks	Not acceptable (R1)	Presumption Against (R2)	Acceptable (R3/4) (with interceptor)
- garage forecourts	Not acceptable (R1)	Presumption Against (R2)	Acceptable (R4) (with interceptor)
- major roads	Not acceptable (R1)	Presumption Against (R2). Acceptable only in exceptional circumstances	Acceptable only if investigation favourable and with adequate precautions (R4)
Industrial Sites	Not acceptable (R1)	Presumption Against (R2)	Acceptable only if investigation favourable and with adequate precautions (R3/4)
	RESOURCE PROTECTION		
Activity	Major Aquifer	Minor Aquifer	Non-Aquifer
C. DISCHARGES OF SURFACE WATER TO SOAKAWAY FROM:			
Roof Drainage	No objection (R5)	No objection (R5)	No objection (R5)
Impermeable Areas			
- public/amenity	Acceptable (R4)	Acceptable (R4)	Acceptable (R4)
- large car parks	Acceptable (R4) (with interceptor)	Acceptable (R4) (with interceptor)	Acceptable (R4) (with interceptor)
- lorry parks	Acceptable (R4) (with interceptor)	Acceptable (R4) (with interceptor)	Acceptable (R4) (with interceptor)
- garage forecourts	Acceptable (R4) (with interceptor)	Acceptable (R4) (with interceptor)	Acceptable (R4) (with interceptor)
- major roads	Acceptable (R4) (subject to investigation and with interceptor)	Acceptable (R4) (subject to investigation and with interceptor)	Acceptable (R4) (with interceptor)
Industrial Sites	Acceptable only if investigation favourable and with adequate precautions (R3/4)	Acceptable (R4) (subject to investigation) with interceptor	Acceptable (R4) (subject to investigation) with interceptor

R1 - Prohibit/object in principle
R2 - Presumption against
R3 - Prohibition notice/Consent to discharge
R4 - No objection subject to standard conditions
R5 - No objection

* See definitions in Section 4.6.3

Detailed information on the assessment of vulnerability, and the key factors which affect it, are given in the NRA's policy document (NRA, 1992). The NRA is in the process of compiling a series of maps of groundwater vulnerability for the UK, by reference to geological conditions and soil characteristics. Each map will be made available on completion. Around each groundwater abstraction borehole, the NRA has defined three source protection zones, as follows:

- zone I (inner zone): the area defined by a minimum of 50 m radius, or the distance corresponding to a 50 day travel time from any point below the water table, to the point of abstraction

- zone II (outer zone): similar to the inner zone (I), with a 400 day travel time and or a minimum of 25% of the source recharge area, whichever is the larger.

- zone III (catchment zone): includes the whole catchment area for the source.

Table 4.9 shows the NRA's acceptability matrix for discharges to underground strata. In this matrix stormwater is classified with respect to the type of surface drained. The NRA's views regarding the acceptability of infiltrating the stormwater from each category, in both source and resource protection zones, are given. The matrix is useful as an initial guide when considering the feasibility of an infiltration system. It should be used in conjunction with the GPP document.

If the source of the stormwater is solely roof surfaces there are unlikely to be any requirements relating to water quality. If the surface that is being drained could be contaminated with oils or petrol, for example car parks or roads, then an interceptor may need to be included. For anything other than roof drainage, wherever possible physical or biological methods should be used to improve water quality. The selection of a physical or a biological method will depend upon the nature of the site and the arrangements for maintenance. For the use of infiltration devices for major roads, quality will be an important issue to which special consideration will have to be given, particularly given the risk of accidental spillages onto the road surface. Industrial sites will always require special investigation because of the wide range of possible pollutants and potentially high concentrations involved. The suitability of using infiltration will depend upon the type of industry present.

It is emphasised that an infiltration device will normally have to comply with the NRA policy on discharges to underground strata. Discharge consent would normally be refused if List 1 substances are present (EC Groundwater Directive). It should be noted that, except for roof drainage water, the NRA are unlikely to approve the use of infiltration systems close to water supply boreholes in source protection Zone 1.

4.6.4 A checklist for water quality

The water quality aspect of infiltration drainage systems is an important factor to be considered during the initial planning stages of a proposed urban drainage scheme. Uncertainties as to the effect of stormwater infiltration on the quality of the receiving groundwater, mean that decisions regarding pollution control should be made with caution.

The following checklist highlights the main points to be considered when planning to construct an infiltration system for the disposal of stormwater:

1. Categorise the stormwater source area, i.e. the type of surface over which runoff flows to the infiltration system.

2. Assess the likely pollutant content of the stormwater, types and concentrations, if possible from the analysis of stormwater samples.

3. Carry out a hydrogeological assessment of possible impact on groundwater (see CIRIA, 1994).

4. Check the NRA policy on discharges to underground strata for the relevant stormwater source(s), and if runoff is from any other source than roof surfaces, consult the NRA about the vulnerability of groundwater in the area, and the sort of pollution controls which may be required.

5. Decide on the type and basic design of infiltration system.

6. Consider pollution control methods which are appropriate for:

   - the pollutants to be removed
   - the size of source catchment or volume of stormwater runoff
   - the type of infiltration device.

   The following general points should be noted:

   - Stormwater which could contain oils or grease, such as drainage from roads, car parks and garages, should pass through an appropriate interceptor system before being infiltrated. Runoff from major highways may also require provision for the storage of any accidental spillage.

   - Roof drainage is unlikely to require treatment before infiltration. It is still advisable, however, to provide a sediment trap and care should be taken to minimise the use of polluting materials in roof construction.

   - Infiltration of urban stormwater from residential or commercial areas may be given approval only subject to a satisfactory investigation, and the use of one or more physical or biological pollution mitigation measures. The level of treatment required will depend on the vulnerability of the groundwater in the area.

   - Due to the variability of stormwater quality from different industrial sites, infiltration of stormwater from industrial areas will only be acceptable to the NRA subject to investigation and with adequate treatment measures. Infiltration will generally not be acceptable close to a groundwater source borehole.

7. Approach manufacturers and/or consultants where necessary, for design details of pollution removal structures or methods.

8. Following the construction of the infiltration system and any associated pollution mitigation measures, it is recommended that the system is monitored and regularly maintained. If possible, data should be collected which would allow the effectiveness of any water treatment methods to be assessed.

4.7 MAINTENANCE RESPONSIBILITIES

Every infiltration system will require maintenance during its lifetime. One of the factors affecting the useful life of infiltration systems is the silting of the surrounding soil due to the entry of sediment through the system. After prolonged use, a zone of less permeable ground may build up around the system, which leads to a reduction in the discharge capacity and an increase in local pore water pressures. The time taken for a substantial loss of performance to occur depends upon the nature of the soil and the quantity of sediment that enters the system. Though there are a number of systems which have operated satisfactorily for long periods without any maintenance, surface infiltration systems will normally require maintenance of their surface. For example, the vegetation of infiltration basins may require regular cutting and scarifying, and the soil and vegetation may also require replacement at infrequent intervals, while porous pavements may require periodic cleaning to maintain infiltration rates. For vegetated infiltration systems the action of plants and animals may reduce or prevent the clogging of the soil. Where feasible, steps should be taken to prevent clogging of the surrounding soil. Sediment may be removed before entry by the use of traps, ponds or settling basins upstream or sediment may be prevented from entering the surrounding soil by the use of geotextiles.

Sub-surface infiltration systems also require maintenance. In general their maintenance is more difficult than for surface systems because of the problems of access to the infiltrating surfaces.

Maintenance will normally be the responsibility of the owner of the land, unless other legal provisions have been made. It may be possible to arrange for the maintenance of the system by an appropriate sewerage undertaker as part of a commercial arrangement. If adoption is being contemplated, it is recommended that approaches are made to the appropriate authority as early as possible. During the design of the system, advice should be sought from the appropriate authority to establish any specifications which the authority might have to ensure its acceptability for subsequent maintenance. It is the policy of the Statutory Sewerage Undertakers in England and Wales not to adopt infiltration systems.

If the infiltration system provides highway drainage, it may be possible to arrange for it to become the responsibility of the highway authority. The same issues are relevant as those for maintenance discussed above. The highway authority should be approached as early as possible to establish whether it would contemplate adopting the road as a highway and accepting responsibility for the infiltration system, and if so, whether the authority has any specifications which have to be met by infiltration systems proposed for adoption.

Similarly, if the infiltration system is to be incorporated into an area of public open space within the development the local authority may agree to adopt it. Similar approaches should be made to the authority to ensure that any essential features are incorporated into the design.

Failure of an infiltration system may render the owner liable for any resulting damage arising from problems of quantity, quality or of a geotechnical nature.

5 Design considerations for particular systems

5.1 INTRODUCTION

In this Section issues relating to infiltration basins and swales, infiltration trenches, porous or infiltration pavements, infiltration blankets and soakaways are described. There is a separate subsection for each type of infiltration system. A subsection on a particular type of infiltration system should be read in conjunction with Section 4 which contains general considerations applicable to all types of system. For examples of existing schemes the reader should refer to CIRIA Project Report 22 (CIRIA 1995) which contains useful details of a range of infiltration systems.

5.2 INFILTRATION BASINS AND SWALES

5.2.1 Description

a) Infiltration basins

An infiltration basin is an area of land surrounded by a bank or berm, which retains stormwater until it has infiltrated through the base and sides of the basin (see Figure 2.4). The basin is frequently excavated in the ground surface, but occasions do occur where berms are used to enclose an area of land on the ground surface, or on one side where the basin is constructed on sloping ground. If there is concern about the impact of large storm events infiltration basins may be provided with overflows which discharge to another means of drainage. In such circumstances the basin acts partly to infiltrate the flow and partly to attenuate the peak by providing detention storage thereby providing some protection against extreme storm events.

b) Swales

A swale is a grass-lined channel with shallow side slopes (see Figure 2.5). Infiltration will occur along its length and it will also act as conveyance channel, storing and discharging water to other forms of drainage. The infiltration properties may be enhanced by the provision of low dams across the swale to divide it up into a series of pounds to encourage detention storage and infiltration. This becomes necessary on slopes steeper than 4%. The shallow side slopes of swales mean that vegetation can be controlled using machines, such as machine-mowers, for example. Swales can provide a low-cost, aesthetic means of both infiltrating and transporting storm drainage while helping to attenuate the flow.

Many of the issues relating to construction and management are similar for both infiltration basins and swales and so the two are considered together. If check dams are placed across swales to prevent flow the hydraulic design method becomes identical with that for an infiltration basin. There is a difference, however, related to the fact that a swale has both a conveyance and an infiltration function. A swale is primarily a channel, often with a detention storage function, which may be used as an infiltration system and thus an important element in the hydraulic design of a swale is its conveyance. The hydraulic design method presented in this manual applies only to infiltration systems and not to channels. To design the conveyance aspects of swales the reader is referred to standard texts on flow in channels.

5.2.2 Suitability of site

On aesthetic grounds and to enhance infiltration it is usual for infiltration basins and swales to be vegetated. The characteristics of the soil can affect the establishment and growth of vegetation. If there is any concern about the suitability of the site for the required vegetation then these characteristics should be investigated.

The environmental requirements relevant to plant growth are:

- the nature of the soil
- the bio-climate.

Nature of the soil

The important soil characteristics are:

- texture
- soil structure
- rooting potential
- soil water capacity
- permeability and water acceptance
- ion exchange capacity
- ease of cultivation.

The physical, water and chemical characteristics of the soil can be combined into an overall assessment of soil potential for plant growth. A scheme for this is given in Table 5.1.

Class A is the highest quality and suitable for situations where good quality fertile topsoil is necessary. However, class C, whilst of poorer quality, would be suitable for many situations. In many cases it would be possible to modify or manage a class B or C soil to improve its quality.

Bio-climate

The climatic factors that influence plant behaviour and performance are:

- growing season
- moisture regime
- exposure
- seasonal rainfall.

The main parameters defining the bio-climate are:

- rainfall
- potential evapotranspiration
- rainfall erosivity
- growing season
- exposure
- temperature.

Table 5.1 Assessment of soil potential for plant growth, (from Coppin and Richards (1990))

Parameter	Unit	Suitability class			Unsuitable
		A	B	C	
Soil type					
Texture	description[1] and clay %	fLS,SL SZL,ZL	SCL,CL, ZCL,LS	C<45% SC,ZC,S	C>45%
Stoniness	% vol	<5	5-10	10-15	>15
Available water capacity (at packing density 1.4-1.75)	% vol	>20	15-20	10-15	<10
pH		5.7-7.0	5.2-5.5 7.0-7.3	4.7-5.5 7.3-7.8	<4.7 >7.8
Conductivity	mmho/cm	<4	4-5	8.10	>10
Pyrite	% weight	-	<0.2	0.2-3.0	>3.0
Soil fertility					
Total nitrogen	% weight	>0.2	0.05-0.2	<0.5	
Total phosphorus	mg/kg	>37	27-37	<27	
Total potash	mg/kg	>360	180-360	<180	
Available phosphorus	mg/kg	>20	14-20	<14	
Available potassium	mg/kg	>185	90-185	<90	

Notes: 1, f = fine, S = sand, C = clay, L = loam, Z = silt

The overall bio-climate may be derived from statistics available from the Meteorological Office, the Ministry of Agriculture Fisheries and Food and the Soil Survey of England and Wales. Local modifications due to slope and aspect may also have to be applied to this information. A scheme for assessing site potential is given in Coppin and Richards (1990).

5.2.3 Hydraulic design

As there can normally be a significant amount of infiltration through the sides of an infiltration basin or swale, the hydraulic design should be based on the 3-D infiltration method described in Section 4. Consideration should be given to the impact of storms which exceed the design event.

5.2.4 Construction

When constructing infiltration basins or swales it is important to pay careful attention to the selection, handling and treatment of soils which are to be in the potential root zone of the vegetation. Failure to address these problems can have a major impact on the establishment and growth of vegetation.

The selection, handling and treatment of soils which are to be used within the potential root zone of the vegetation, say within 1m of the final ground surface, should take account of:

- their potential as a medium for plant growth
- continuity between soil layers.

Example of design of infiltration basin (see Procedure in Section 4.4.2)

An infiltration basin is proposed to drain an area of 9000 m² at a site near Southampton. It is to be designed on the basis of 1 in 10 year return period. An on-site infiltration test indicated an infiltration coefficient of 0.225 m/h. The first estimate of the base area A_b, is 10 m × 8 m with side slopes of 1 in 2. It is necessary to determine whether this is a suitable area for the base and to calculate the corresponding maximum water depth.

From Figure 4.3 the rainfall ratio for the Southampton region is r = 0.35 and hence, the following design storm event table can be constructed using the procedure given in Section 4.4.1.

Table 5.2 Rainfall duration and intensity

Storm duration D min	Intensity m/h
10	0.074
15	0.061
30	0.039
60	0.025
120	0.015
240	0.009
360	0.007
600	0.005

An infiltration basin is a 3-D system. If we approximate this by a vertically-sided system then, the appropriate design method is given in Section 4.4.2(e).

Step 1: Divide the observed infiltration coefficient by the factor of safety to determine the effective infiltration coefficient.

If the basin is overtopped then it is considered that no damage or inconvenience will result. By referring to Table 4.6 a factor of safety of 1.5 is selected. The effective infiltration coefficient is given by:

$$q = \frac{0.225}{1.5} = 0.15 \text{ m/h}$$

Step 2: As there is no fill in the basin the porosity, $n = 1.0$

Step 3(i): The area to be drained, A_D, is 9000 m²

Step 3(ii): Confirm suitability of infiltration basin

Step 4(i): The width and length of the base of the basin will be 8 and 10 m, respectively

Step 4(ii): The base area, A_b, of the basin is 8 × 10 = 80 m² and the perimeter P is 2 × 8 + 2 × 10 = 36 m

Step 4(iii): $b = \dfrac{Pq}{A_b n} = \dfrac{36 \times 0.15}{80 \times 1} = 0.067 \; h^{-1}$

Steps 5 to 8 Values of h_{max} are found from the equation

$$h_{max} = a \times (\exp.^{(-b \times D)} - 1) \text{ or use Figure 4.5 to determine } h_{max}$$

where $\quad a = \dfrac{A_b}{P} - \dfrac{A_D i}{Pq}$

Table 5.3 presents the results for a range of durations.

Table 5.3 Values of rainfall intensity, duration, b, a and h_{max}

Intensity i(m/h)	Duration D(h)	a	b	h_{max} (m)
0.061	0.25	−99.4	0.067	1.65
0.039	0.50	−62.78	0.067	2.07
0.025	1.00	−39.44	0.067	2.55
0.015	2.0	−22.78	0.067	2.86

Step 8: Select the largest value of h_{max} from Table 5.3. The largest value of h_{max} is 2.9 m. The depth of the basin would have to be at least 2.9 m which was considered unsatisfactory so the proposed base area, A_b, was increased to 20 m × 8 m.
Steps 4 to 8 of the hydraulic calculations have then to be repeated with the revised value of the base area.

Repeat Step 4(ii) $A_b = 160$ m^2
$\qquad\qquad\qquad P = 56$ m

Repeat Steps 5 to 8

Table 5.4 Values of rainfall intensity, duration, b, a and h_{max}

Intensity i(m/h)	Duration D(h)	a	b	h_{max} (m)
0.061	0.25	−62.50	0.0525	0.81
0.039	0.50	−38.93	0.0525	1.01
0.025	1.00	−23.93	0.0525	1.22
0.015	2.00	−13.21	0.0525	1.32
0.009	4.00	−6.786	0.0525	1.28

Repeat Step 8 from Table 5.4 the maximum depth of water in a basin of size 20 m × 8 m would be 1.32 m.

Time for emptying

Since it is a 3-D infiltration system, the time for emptying from full to half-full is

$$t = \frac{nA_b}{qP} \log_e \left[\frac{h_{max} + \frac{A_b}{P}}{\frac{h_{max}}{2} + \frac{A_b}{P}} \right]$$

$$= \frac{1 \times 160}{0.15 \times 56} \log_e \left[\frac{1.32 + \frac{160}{56}}{\frac{1.32}{2} + \frac{160}{56}} \right]$$

$$= 3.27 \text{ h}$$

The above design method assumed that the sides of the infiltration basin were vertical. For a more sophisticated method of analysis which takes account of sloping sides see Bettess *et al* (1995).

For an alternative approach to the design of a swale from the viewpoint of its behaviour as an infiltration system see CIRIA Report 124, Vol 3, Section 5.

Irreparable damage can be done in terms of potential for plant growth and infiltration capacity if a soil is not handled correctly. The following practices should be avoided:

- stockpiling topsoil in such a way that the natural aerobic soil organisms are destroyed
- destroying the existing soil structure
- compacting soil to excessive densities so that infiltration is reduced and root growth inhibited.

To reduce the risk of damage to the soil the following practices should be followed:

- where possible, excavators and dumpers, rather than scrapers, should be used
- double handling should be avoided
- stockpiles, where they are necessary, should be shallow and not heavily compacted
- any handling or tracking over soil should only occur when compatible with the soil moisture conditions
- indiscriminate tracking by heavy, earthmoving machinery should be avoided.

If excessive soil compaction has taken place deep cultivation can be used to:

- increase infiltration into the soil
- reduce soil density
- provide drainage pathways through the soil.

The ideal implement to use for deep cultivation is a winged tine, working to 0.6 to 0.8 m depth with the tines spaced 1.0 to 1.2 m apart.

If seeding is to take place, the final surface should be prepared accordingly, for example by :

- drilling: a reasonably level surface and a loose tilth are necessary for the seed-drill machine
- broadcast and harrowing: a reasonably fine tilth is needed to allow the harrows to bury the seed to the correct depth
- broadcast without harrowing, including hydroseeding: a rough surface texture will result in the best seedling establishment.

Information on appropriate seed mixtures can normally be obtained from suitable seed merchants.

For further information see Coppin and Richards (1990).

Where flow enters the basin or swale from a pipe or channel, there is the potential for scour to occur. To prevent such scour, stone or other protection should be placed around the discharge point. To reduce soil erosion the longitudinal slope of swales should be less than 5%.

5.2.5 Selection of vegetation type

(a) Selection criteria

The selection of the appropriate vegetation depends upon:

- ecological or site conditions
- method of propagation
- management.

(b) Ecological or site conditions

The specific tolerances and preferences of individual plant species are widely available in the literature (e.g. Coppin and Bradshaw, 1982; Hubbard, 1986).

The bio-climate of a particular site is fixed and species selection should take into account their tolerance to factors such as cold, exposure or short growing season.

Soil fertility is a significant parameter but it may be modified by the use of fertilisers or by using species such as clovers, vetches or lupin, which fix nitrogen.

The soil water regime will influence the selection of suitable vegetation types. Because of the particular application to infiltration basins, plants should be tolerant of wet conditions and short periods of inundation.

(c) Plant community

It is unlikely that one species alone will satisfy all requirements. It is advisable, therefore, to use a mixture of complementary species whose balance will adjust in response to different environmental conditions. Grasses are usually selected as the main component of groundcover and are often mixed with legume or other herbaceous species. The use of low-maintenance mixtures which have been developed for shortness of growth may be considered.

(d) Method of propagation

The most widely used form of propagation is from seeds. Seeds can be cheap and are useable in a wide range of situations, but during the establishment phase they are very vulnerable to desiccation and predation.

(e) Management

In selecting appropriate types of vegetation the future management of the system should be considered (see Section 4.7).

For further information on this subject the reader is referred to Coppin and Richards (1990).

5.2.6 Maintenance

(a) Inspection

The performance of the system should be checked after any major storms in the first few months after construction. Particular attention should be paid to how long runoff remains in the basin as retention of water for periods in excess of 48 or 72 hours may indicate that the infiltration capacity of the basin has been overestimated. Inspections should look for evidence of sediment deposition, lower spots, excessive soil compaction or poor soils. Where necessary remedial work should be carried out.

Subsequently, the basin should be inspected annually. The inspection should include checks on vegetation condition, tree growth on the embankment, condition of rip-rap at the inlet, sediment accumulation and any outlets, if they are present.

(b) Managing grass and herbaceous swards

Grass is usually maintained by mowing. Frequency can vary from several times a year, to maintain short turf, to very occasional cutting, every one to three years, to prevent shrub invasion. Regular cutting stimulates side shoots in grasses and results in a sward which is thicker and better able to withstand surface abrasion. The frequency and height of cutting should be matched to the purpose and function required of the vegetation.

Grazing is a cost-effective method for controlling vegetation. Sheep are preferred because they are light and have small feet which do not damage the sward.

Fertilisers are important in the management of soil fertility, but advice on their use should always be sought as it is not possible to prescribe levels without assessing each situation individually. The over application of fertilisers should be avoided.

(c) Trash

Trash will tend to collect in basins. It is good practice to remove all debris and litter from time to time.

(d) Deep tilling

Over time the infiltration capacity of the basin may reduce due to clogging of the surface. The infiltration capacity may be restored by carrying out deep tilling to break up the surface layer, followed by re-grading and levelling. Sand or organic material may also be tilled into the basin to improve the infiltration characteristics.

(e) Sediment removal

If sediment accumulates in the basin it should be removed periodically.

5.2.7 Advantages and disadvantages

The main advantages and disadvantages of infiltration basins and swales relative to other infiltration systems are:

Advantages

- can be easily maintained
- can be attractively fitted into the landscape and can enhance the environment
- can be used to drain large areas
- can incorporate biological methods to enhance the water quality
- can allow some types of alternative use of the land
- the performance of the system can be easily monitored.

Disadvantages

- requires a significant area of land
- could restrict some of the alternative uses for the land due to periodic inundation.

5.3 INFILTRATION TRENCHES WITH SURFACE INFLOW

5.3.1 Description

Infiltration trenches are in many respects similar to trench soakaways and consist of a trench filled with permeable material (see Figure 2.8). Sometimes they are constructed along the sides of large impermeable areas to take stormwater runoff. In these cases the entry of water into the system may be distributed along its length. Where the entry of water is confined to a few locations along the length of the trench, the system operates in a similar fashion to a trench soakaway (see Section 5.6).

5.3.2 Suitability of site

The nature of the soil must be suitable for infiltration. Infiltration trenches should not be located close to the foundations of other structures or services (see Section 4.5).

5.3.3 Hydraulic design

As the infiltration is predominantly three-dimensional, the hydraulic design should be based on 3-D infiltration as described in Section 4. An example calculation is given below:

> **Example of design of an infiltration trench (see Procedure in Section 4.4.2)**
>
> A rectangular trench soakaway of 0.8 m × 14.0 m base area and with vertical sides is required to drain an area of 320 m^2 for a site near Southampton. The fill of the trench has a porosity of 0.35. An on-site infiltration test indicated an infiltration coefficient of 0.18 m/h. The maximum depth of water is to be calculated.
>
> As the infiltration trench is a vertical-sided infiltration system, the procedure described in Section 4.2.2(e) is used with equation 8

Step 1: Divide the observed infiltration coefficient by the factor of safety to determine the effective infiltration coefficient. From Table 4.6 the appropriate factor of safety is 1.5. The effective infiltration coefficient is given by

$$q = \frac{0.18}{1.5} = 0.12 \text{m/h}$$

Step 2: Porosity of fill, $n = 0.35$

Step 3(i): $A_D = 320 \text{ m}^2$

Step 3(ii): An infiltration trench is to be used

Step 4(i): The proposed dimensions are $W = 0.8$ m and $L = 14.0$ m

Step 4(ii): Base area, $A_b = 0.8 \times 14.0 = 11.2 \text{ } m^2$, and perimeter,

$$P = 2 \times (0.8 + 14.0) = 29.6 \text{ m}$$

Step 4(iii): $b = \frac{Pq}{A_b n} = \frac{29.6 \times 0.12}{11.2 \times 0.35} = 0.906 \text{ h}^{-1}$

Steps 5 to 8: Values of h_{max} are found from the equation

$$h_{max} = a \times \left(\exp.^{(-b \times D)} - 1\right), \text{ where } a = \frac{A_b}{P} - \frac{A_D i}{Pq}.$$

For example when $D = 0.25$ and $i = 0.061$,

$$a = \frac{11.2}{29.6} - \frac{320 \times 0.061}{29.6 \times 0.12} = -5.117 \text{ and}$$

$$h_{max} = -5.117 \times \left(\exp.^{(-0.906 \times 0.25)} - 1\right) = 1.04 \text{ m} \quad \text{or}$$

else h_{max} may be determined from Figure 4.5.

Table 5.5 presents the results for a range of durations.

Table 5.5 Values of rainfall intensity, duration, b, a and h_{max} for an infiltration trench

Intensity i (m/h)	Duration D (h)	b	a	h_{max} (m)
0.061	0.25	0.906	−5.117	1.04
0.039	0.50	0.906	−3.135	1.14
0.025	1.00	0.906	−1.874	1.12
0.015	2.00	0.906	−0.973	0.81

Step 8: Select the largest value of h_{max} from Table 5.5. The largest value of h_{max} is 1.14 m, which occurs during a $M10$ 0.5 hour storm. The depth of trench required is, therefore, at least 1.14 m.

Step 9: This value would appear to be satisfactory. Should a deeper trench be preferred then it would be possible to shorten the length or decrease the width of the trench.

Time for emptying

The method to determine the time taken for the infiltration trench to drain down to half-full may be found in Section 4.2.2(f)

The time taken to half-empty the trench is given by

$$\frac{A_b n}{Pq} \log_e \left[\frac{h_{max} + \frac{A_b}{P}}{\frac{h_{max}}{2} + \frac{A_b}{P}} \right]$$

$$= \frac{0.35 \times 11.2}{0.12 \times 29.6} \log_e \left[\frac{1.14 + \frac{11.2}{29.6}}{0.57 + \frac{11.2}{29.6}} \right] = 0.52 \log_e$$

5.3.4 Construction

Infiltration trenches normally consist of an excavated trench back-filled with suitable granular material. In the hydraulic calculations, the porosity of the fill should be taken into account.

If the entry of water is confined to a few locations along the length of the trench a perforated or porous distributor pipe should be provided.

The base of the trench should be horizontal.

A minimum of two inspection access points should be provided, though the number may be increased depending upon the length of the trench.

If water enters from a drain discharge point it should be fitted with a sediment trap and an petrol/oil interceptor, if appropriate. In the case of distributed flow entry into the trench, there is no opportunity for the use of grit traps or interceptors. In these cases, therefore, one should be confident that their absence will not adversely affect the useful life of the system or water quality. Where the flow entry is distributed it may be possible to use a grassed zone as a buffer to trap sediments. Such a buffer zone should extend approximately 5m from the edge of the trench. In order to prevent soil being washed from the buffer zone into the trench, the buffer zone should be turfed or some other form of erosion protection used, e.g. pre-seeded geo-fabrics.

5.3.5 Maintenance

Any sediment traps or petrol.oil interceptors should be inspected regularly, at least once a year being recommended. Any cleaning required, identified from inspection, should be undertaken without delay.

The trench should be inspected several times in the first few months of operation and then annually thereafter. These inspections should be conducted after large storms to check for surface ponding or other indications of reduced infiltration.

If a grass buffer strip is used, this should be inspected annually. Growth should be vigorous and dense. Bare spots, eroded areas or damaged areas should be re-seeded or turfed. Such grass strips should be mown periodically, though the filter performance will be reduced if the grass is cut too short.

5.3.6 Advantages and disadvantages

The main advantages and disadvantages of infiltration trenches relative to other infiltration systems are:

Advantages

- visually unobtrusive
- allows use of site for other activities
- can be fitted into what would otherwise be non-utilised areas.

Disadvantages

- no opportunity for biological treatment of stormwater
- long-term maintenance is difficult: if performance deteriorates with time the whole structure may have to be replaced
- difficult to monitor performance.

5.4 INFILTRATION PAVEMENTS

5.4.1 Description

An infiltration, or porous, pavement is a pavement which is constructed to allow infiltration through the surface and into the sub-grade below (see Figure 2.3). The water may then discharge to the soil beneath or be collected and discharged to positive drainage. The most frequent use is for car parking areas. There are a range of suitable pavement materials but the most commonly used are porous macadam and concrete blocks set in a granular fill.

This section addresses the use of porous pavements in conjunction with infiltration. It does not address the use of such pavements in conjunction with positive drainage.

A typical cross-section consists of 50 to 100 mm of porous macadam overlying a layer of porous fill material. The runoff infiltrates through the pores of the macadam into the void spaces of the underlying fill material and thus into the underlying soils. The storage capacity of the system is primarily a function of the depth of the underlying fill material. If the pavement is to be used for vehicles then this fill material has a dual role of acting both as a storage medium for stormwater and as a sub-base for the surface pavement. The depth and design of this sub-layer should be sufficient to satisfy both roles.

A potential problem with some types of infiltration pavements is that if the surface becomes clogged it can be difficult and costly to rehabilitate. There are many types of pavements for which this is not a problem, for example, concrete block systems. See CIRIA Project Report 22 (CIRIA 1995) which provides details of case studies.

5.4.2 Suitability of site

The nature of the soil must be suitable for infiltration. Infiltration pavements should not be located close to the foundations of other structures or services (see Section 4.5).

5.4.3 Hydraulic design

As the flow is predominantly from the base of the system the hydraulic design should be based on plane infiltration as described in Section 4. An example calculation is given below:

Example of design of an infiltration pavement (See Procedure in Section 4.4.2)

An infiltration pavement is to be designed for a site near Southampton. A sub-grade fill of 0.1 m depth is to be used which has a porosity of 30%. A soakage test has indicated that the infiltration rate is 0.25 m/h. The maximum allowed depth of water in the sub-base of the pavement is 0.10 m. It is necessary to confirm that this depth of water will not be exceeded during a 10-year storm event.

Since the site is located close to Southampton, the rainfall statistics used are the same as those in the previous example.

As we are designing a plane infiltration system the procedure described in Section 4.4.2(d) is used, but with equation 5 for an infiltration pavement.

Step 1: Divide the observed infiltration coefficient by the factor of safety to determine the effective infiltration coefficient. If the pavement floods minor inconvenience will result. From Table 4.6, the appropriate factor of safety is, therefore, 5. The effective infiltration coefficient is given by:

$$q = \frac{0.25}{5} = 0.05 \text{ m/h}$$

Step 2: The porosity of fill material is given as 30% so $n = 0.3$

Step 3: is omitted because $A_D = A_b$, therefore $R = 1$

Steps 4 to 6: For values of rainfall intensity and duration, the corresponding values of h_{max} are calculated in Table 5.6 using the formula

$$h_{max} = \frac{D}{n}(i-q)$$

Table 5.6 Values of rainfall intensity, duration and highest water level, h_{max}, for an infiltration pavement

Intensity i (m/h)	Duration D (h)	h_{max} (m)
0.10	0.083	0.014
0.074	0.167	0.013
0.061	0.25	0.009
0.039	0.50	− 0.0018

Step 7: From Table 5.6 select the largest value of h_{max}. The largest value of h_{max} is 0.014 m.

The calculations show that the maximum allowed depth of 0.10 m will not be exceeded. It should be noted that the negative value for h_{max} for the storm of duration 0.5 h indicates that, for this storm, the infiltration capacity of the pavement exceeds the rainfall intensity, that is, for this storm event the water would infiltrate from the pavement as quickly as the rain falls onto the surface. High water levels are only likely to occur in the system during short sharp rainfall events.

Time for emptying

The time taken for the system to half-empty is given in Section 4.4.2(f) by

$$\frac{nh_{max}}{q} = \frac{0.3 \times 0.1}{2 \times 0.05} = 0.3 \text{ hour}$$

Alternatively, the minimum feasible infiltration coefficient can be calculated on the grounds of half-emptying time, for the proposed design. From Section 4.4.2(f) and taking a design h_{max} of 0.1 m,

$$q = \frac{0.3 \times 0.1}{48} = 0.0006 \text{ m/h}$$

As this minimum value is considerably less than the design infiltration coefficient of 0.05 m/h, there are no concerns over the half-emptying time.

5.4.4 Construction

Infiltration pavements are normally used to provide a surface which can be used for activities such as car parking or as a yard. The construction of the pavement should be sufficient to withstand this use. Typical surfaces that have been used include permeable tarmacadam and concrete blocks with the spaces between filled with soil or gravel. The surface should be capable of high infiltration rates allowing for the fact that these may reduce markedly with time. Underlying the surfacing, a layer of fill material is required to provide storage during a rainfall event. Depending upon the nature of the fill and the overlying surface material, a filter or geotextile may be required to separate the materials.

5.4.5 Maintenance

It has been suggested by some (see CIRIA Project Report 21 (CIRIA 1995)) that maintenance personnel should diligently clean dirt and other matter from the surface to minimise clogging of the surface layers, although an annual sweeping may be the norm. It has been reported that, although asphaltic materials tend to knead to a tighter, less permeable condition under traffic, this action may be nullified by the formation of cracks and open joints which maintain the surface infiltration rate (Cedergen, 1974). Other experience with porous macadam surfaces has shown neither a tendency to knead nor the development of cracks. Where the surface consists of a lattice of blocks it is important to maintain the infiltration capacity of the material between the blocks. This is aided if the infiltration surface is set slightly below the load-bearing surface (see Figure 5.1). This allows the growth of vegetation which may be beneficial both aesthetically and in maintaining the infiltration capacity. If traffic has direct contact with the infiltration surface then vegetation growth may be inhibited and the infiltration surface may become compacted and clog (Escritt and Howarth, 1984).

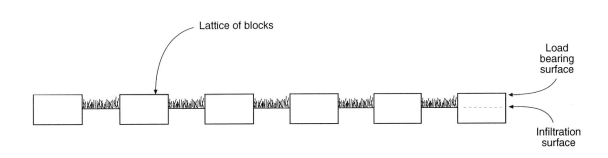

Figure 5.1 Lattice of blocks with infiltration surface set below load bearding surface

5.4.6 Advantages and disadvantages

The main advantages and disadvantages of infiltration pavements relative to other infiltration systems are:

Advantages

- requires no extra land other than that which is already allocated to car parking or yards
- may provide better skid resistance and reduced spray than a standard macadam surface.

Disadvantages

- it is not possible to intercept oils and petrols before the runoff enters the infiltration system
- silt removal prior to infiltration is not possible and so surface may be liable to premature clogging
- depending upon the nature of the surface, once clogged it may be difficult and costly to rehabilitate.

5.5 INFILTRATION BLANKETS

5.5.1 Description

Infiltration blankets consist of a porous granular material overlying the infiltration surface and which itself lies under the ground surface (see Figure 2.9). The infiltration blanket may, therefore, be overlain by material unrelated to the infiltration system. Thus the ground surface may be used for many activities while not affecting the underlying infiltration system. Infiltration blankets are normally associated with the disposal of water from a point source, such as a traditional pipe system.

5.5.2 Suitability of site

The nature of the soil must be suitable for infiltration. Infiltration blankets should not be located close to the foundations of other structures or services (see Section 4.5).

5.5.3 Hydraulic design

As the flow is predominantly from the base of the system the hydraulic design should be based on plane infiltration as described in Section 4. An example calculation is given below:

5.5.4 Construction

The infiltration blanket should be horizontal. The porous fill material forming the storage medium for the stormwater should be separated from the overlying material. A permeable geotextile is normally appropriate. Similarly the fill material should also be separated from the underlying soil using a geotextile.

If the incoming stormwater comes from a point source, such as a traditional piped system, a system of distributor pipes is normally required within the blanket. These should consist of perforated or porous pipes.

Example of design of an infiltration blanket (see Procedure in Section 4.4.2)

In an area close to Southampton, an infiltration blanket is required to drain an area of 4000 m^2. For the purposes of this example it is assumed that the maximum allowed depth of water in the blanket is 0.10 m. (It should be noted that there is no intrinsic reason why, in other examples, deeper water levels would not be acceptable in an infiltration blanket). An on-site infiltration test has indicated that the infiltration coefficient is 0.15 m/h. The porosity of the fill is 0.3.

We have, therefore: A_D = 4000 m^2 and n = 0.3

As it is plane infiltration we use equation (4) from Section 4.2.2(d) and again follow the procedure described in that section:

Step 1: Divide the observed infiltration coefficient by the factor of safety to determine the effective infiltration coefficient. If the blanket floods no damage or inconvenience will result. From Table 4.6 the appropriate factor of safety is, therefore, 1.5.
The effective infiltration coefficient is given by

$$q = \frac{0.15}{1.5} = 0.10 \text{m/h}$$

Step 2: The porosity, n = 0.3

Step 3(i): The area to be drained, A_D, is 4000 m^2. It is proposed to use an infiltration blanket with an infiltration surface area, A_b, of 800 m^2

Step 3(ii): The drainage ratio, R, is therefore $\dfrac{A_D}{A_b} = \dfrac{4000}{800} = 5$

Steps 4 to 6: Table 5.7 shows the values of h_{max} calculated for a range of rainfall durations using the formula,

$$h_{max} = \frac{D}{n}(iR - q)$$

Table 5.7 Values of rainfall intensity, duration and highest water level, h_{max}, for an infiltration blanket

A_b (m^2)	Intensity i (m/h)	Duration D (h)	h_{max} (m)
800	0.074	0.167	0.150
	0.061	0.25	0.176
	0.039	0.50	0.166
	0.025	1.00	0.100

Step 7: Select the largest value of h_{max} from Table 5.7. The largest value of h_{max} is 0.176 m.

As h_{max} is larger than the maximum allowable, it would suggest that a larger infiltration blanket is required. The designer may select a larger size and return to step 3. Alternatively, the minimum infiltration area may be found for a set h_{max} by using the procedure described in Section 4.4.2(d) and equation 7.

Steps 1 and 2: These steps have already been conducted in the procedure above.

Step 3(i): The area to be drained, A_D, is 4000 m²

Step 3(ii): The highest allowable water level, h_{max}, is 0.1 m

Step 4 to 6: Table 5.8 shows the area required to contain the rainfall events for a range of rainfall durations, using the formula

$$A_b = \frac{A_D \, i \, D}{n h_{max} + q D}$$

Table 5.8 Values of rainfall intensity, duration and infiltration area, A_b, for an infiltration blanket

h_{max} (m)	Intensity i (m/h)	Duration D (h)	A_b (m²)
0.1	0.074	0.167	1060
	0.061	0.25	1110
	0.039	0.50	975
	0.025	1.00	770

Step 7: Select the largest value of A_b from Table 5.8. The infiltration blanket will meet the design criteria if an infiltration surface area of at least 1200 m² is used.

Time for emptying

The minimum feasible infiltration coefficient is the same as for the infiltration pavement example above. Alternatively, the time for emptying from full to half-full is

$$t = \frac{n h_{max}}{2q}$$

$$= \frac{0.3 \times 0.1}{2 \times 0.1} = 0.15 \text{ hour}$$

The pipe system upstream of the infiltration blanket should be fitted with sediment traps and, where appropriate, petrol/oil interceptors.

5.5.5 Maintenance

Depending upon the nature of the overlying material, access to the infiltration system and hence maintenance can be difficult. Any sediment traps or oil interceptors should be cleaned regularly. The frequency of such cleaning will depend upon the location and the size and type of area drained. The minimum frequency should be once a year.

5.5.6 Advantages and disadvantages

The main advantages and disadvantages of infiltration blankets relative to other infiltration systems are:

Advantages

- usually unobtrusive
- allows use of site for other surface activities.

Disadvantages

- maintenance of the blanket is difficult. If performance deteriorates with time then the whole structure may have to be replaced. Regular maintenance of sediment traps and any petrol/oil interceptors upstream can significantly reduce the problem
- difficult to monitor performance.

5.6 SOAKAWAYS INCLUDING TRENCH AND LINKED SOAKAWAYS

5.6.1 Description

Soakaways

A soakaway is a below ground system capable of disposing of the discharge from a piped drainage system. It may consist, for example, of a clean stone-filled excavation or be constructed from pre-cast, perforated concrete rings (see Figures 2.6 and 2.7). During a storm event some water is stored in the soakaway while infiltration takes place through the sides and base. In essence soakaways differ from infiltration trenches only in shape, but in practice soakaways come in many forms, and may be constructed in a number of different ways.

Linked soakaways

Depending upon the size of the area to be drained and the area of the site, linked soakaways may be used, that is, two or more soakaways joined together by connecting pipes. These normally provide a larger infiltration area for a given storage volume than a single soakaway.

Such soakaways may be linked either in series or in parallel. In any particular situation the configuration of the site normally dictates which arrangement is adopted.

Trench soakaways

Trench soakaways are similar in certain aspects to infiltration trenches. If the site allows the use of trench soakaways then for a given volume of excavation they may provide better infiltration characteristics than more traditional cylindrical soakaways.

5.6.2 Suitability of site

The nature of the soil must be suitable for infiltration. Soakaways should not be located close to the foundations of other structures or services (see Section 4.5).

5.6.3 Hydraulic design

As the infiltration is from the base and sides of the system the hydraulic design should be based on 3-D infiltration as described in Section 4. An example calculation is given below:

Example of hydraulic design of a soakaway (see Procedure in Section 4.4.2)

A soakaway is required to drain an area of approximately 200 m². The soakaway will be filled with a granular material of porosity 0.35. An in-situ infiltration test has indicated that the infiltration coefficient is 0.45 m/h. The soakaway is to be 2.0 m square.

As the soakaway is a vertical-sided infiltration system, the procedure described in Section 4.2.2(e) is used with equation 8.

Step 1: Divide the observed infiltration coefficient by the factor of safety to determine the effective infiltration coefficient. If the soakaway is overtopped then minor inconvenience will result. From Table 4.6 the appropriate factor of safety is 3. The effective infiltration coefficient is given by:

$$q = \frac{0.45}{3} = 0.15 \text{ m/h}$$

Step 2: Porosity of fill, $n = 0.35$

Step 3(i): $A_D = 200 \text{ m}^2$

Step 3(ii): A rectangular soakaway is to be used

Step 4(i): The proposed base is 2 m × 2 m

Step 4(ii): Base area, $A_b = 2 \times 2 = 4 \text{ m}^2$, and perimeter, $P = 2(2+2) = 8 \text{ m}$

Step 4(iii): $b = \dfrac{8 \times 0.15}{4 \times 0.35} = 0.857 \text{ h}^{-1}$

Steps 5 to 7: Values of h_{max} are found from the equation

$$h_{max} = a \times \left(\exp.^{(-b \times D)} - 1\right), \text{ where } a = \frac{A_b}{P} - \frac{A_D i}{Pq}.$$

For example when $D = 0.25$ and $i = 0.061$,

$$a = \frac{4}{8} - \frac{200 \times 0.061}{8 \times 0.15} = -9.667 \text{ and}$$

$$h_{max} = -9.667 \times \left(\exp.^{(-0.857 \times 0.25)} - 1\right) = 1.86 \text{ m}$$

or else h_{max} may be determined from Figure 4.5.

Table 5.9 presents the results for a range of durations.

Table 5.9 Values of rainfall intensity, duration, b, a and h_{max} for a cylindrical soakaway

Intensity i (m/h)	Duration D (h)	b	a	h_{max} (m)
0.061	0.25	0.856	−9.667	1.86
0.039	0.50	0.856	−6.00	2.09
0.025	1.00	0.856	−3.667	2.11
0.015	2.00	0.856	−2.00	1.64

Step 8: Select the largest value of h_{max} from Table 5.9. The largest value of h_{max} is 2.11 m, which occurs during a $M10$ 1 hour storm. The depth of soakaway required is, therefore, at least 2.11 m.

Step 9: If this value is unacceptably high then the diameter may be increased or the drained area may be reduced by using more soakaways.

Time for emptying

The method to determine the time taken for this soakaway to drain down to half-full may be found in Section 4.2.2(f). The time taken to half-empty is given by

$$\frac{A_b n}{pq} \log_e \left(\frac{h_{max} + \frac{A_b}{p}}{\frac{h_{max}}{2} + \frac{A_b}{p}} \right)$$

In this case, taking $h_{max} = 2.11$ m, the time taken to half-empty is

$$\frac{0.35}{0.15} \times \frac{4}{8} \times \log_e \left(\frac{2.11 + 0.5}{1.055 + 0.5} \right) = 0.62 \text{ h}$$

Example of Hydraulic design of soakaway with perforated concrete ring units

The diameter of the concrete ring units is to be 2.0 m placed in an excavation of 2.6 m diameter. The volume around the ring units is to be filled with stone with a porosity of 0.35.
Equation 2 can be used to find the appropriate porosity for the system. In this case, r' = 1.0m, W = 2.6 m and L = 2.6 m. The appropriate porosity, n', is provided by

$$n' = \frac{(\pi \times 1.0^2) + 0.35 \times ((2.6 \times 2.6) - (\pi \times 1.0^2))}{2.6 \times 2.6} = 0.65$$

The same calculation procedure as is given in the previous example should be followed, but with a value of porosity of 0.65.

BRE Digest 365 also gives a method for the design of soakaways. As the two methods treat the factor of safety in different ways they may not always give similar results.

5.6.4 Construction

Large soakaways may be constructed using pre-cast perforated concrete ring units or brickwork within a previously created excavation. The volume between the structure and the excavation should be back-filled with granular material. It has been recommended that for concrete ring units the width of backfill outside the ring should be at least 300 mm. In carrying out the hydraulic calculations this volume outside the structure can be included in the storage after suitable allowance is made for the porosity of the fill.

Other soakaways may consist of an excavated hole, back-filled with suitable granular material. In the hydraulic calculations, the porosity of the fill should be taken into account.

The inflows to soakaways should be fitted with sediment traps and, where appropriate, oil interceptors.

Trench soakaways should be constructed with a horizontal base. They should have at least two inspection access points. A horizontal perforated or porous distributor pipe should be laid in the top of the granular fill along the length of the trench. It may be convenient to have several drain–discharge points along the length of the trench, each connected to the soakaway via an inspection access chamber.

Where linked soakaways are used the connecting pipe should be capable of discharging the maximum discharge from the drained area without water appearing on the surface. There should be at least 5 m of undisturbed ground between soakaways.

5.6.5 Maintenance

Any sediment traps or oil interceptors should be cleaned regularly. The frequency of such cleaning will depend upon the location and the size and type of area drained. The minimum frequency should be once a year. If the type of soakaway allows, it should be cleaned by airlift once a year. High pressure jetting equipment should only be used after careful consideration of the potential for driving silts and fine material into the soil or surrounding granular material.

5.6.6 Advantages and disadvantages

The main advantages and disadvantages of soakaways relative to other infiltration systems are:

Advantages

- visually unobtrusive
- allows use of site for other surface activities
- requires little land
- easily assembled and constructed.

Disadvantages

- no biological treatment of stormwater
- long-term maintenance is difficult. If performance deteriorates with time the whole structure may have to be replaced
- difficult to monitor performance.

6 Legal and administrative aspects of infiltration drainage

6.1 SUMMARY

This Section seeks to provide a comprehensive and reasonably concise account of the general law of England and Wales as it applies to infiltration drainage facilities. It will be evident that this spans a range of different kinds of legal provisions, diverse regulatory laws of both national and European origin, which may restrict the construction and impose continuing obligations for the operation of an infiltration system, along with private rights and duties which may exist between individuals and give rise to liability where a facility causes some kind of actionable harm.

The nature of many of the conclusions drawn from this study of the legal aspects are attributable to the relative legal novelty of infiltration systems. There are few direct legal references to infiltration systems. In this section, therefore, reliance has had to be placed on the implicit application of broader principles to the specific issues related to the construction or maintenance of infiltration systems. Clearly, this raises the issue of whether more explicit legal provision should be made in relation to this kind of facility.

Broadly, the areas of law which have been found to be of greatest relevance are as follows. Some of the legal issues are complex and reference should be made to the detailed commentary in the report on legal aspects (see CIRIA Project Report 25 (CIRIA 1995)).

The Section reflects the understanding of the legal position as it was in August 1992.

6.2 LEGAL ASPECTS

The main legal provisions which bear upon the design, construction, use and maintenance of infiltration drainage facilities are summarised below.

6.2.1 Pollution of groundwater

Continuing concerns may arise as to the water quality implications of infiltration drainage systems. A range of criminal offences exists in respect of the pollution of 'controlled waters', which includes groundwater, and it is clear that these offences would be committed where a person 'caused or knowingly permitted' a facility of this kind to be used as a means of polluting subsurface waters. A defence to this offence is provided to highway authorities unless a prohibition is imposed upon particular highway drains by the National Rivers Authority. Otherwise, an offence will arise in respect of groundwater pollution by means of an infiltration system unless the entry of polluting matter or effluent is the subject of a discharge consent granted by the Authority.

6.2.2 Protection of groundwater resources against pollution

(a) Water undertaker interest

Another facet of the problem of groundwater contamination, potentially arising through the operation of infiltration facilities, relates to water resources. The obligation upon statutory water undertakers to supply wholesome water to domestic premises is clearly dependent upon the protection of water resources, including aquifers that are used for this purpose. Accordingly, water undertakers have a significant interest in the maintenance of purity of those resources, and the National Rivers Authority is obliged to have regard to the water supply duties imposed upon undertakers in exercising its powers.

(b) National Rivers Authority interest

It follows, therefore, that in regulating the operation of an infiltration facility, through a discharge consent or otherwise, the National Rivers Authority must have regard not only to the general protection of the aquatic environment, but also to the effects which proposed discharges into aquifers may have upon underground water which may subsequently be used for water supply purposes.

(c) NRA Groundwater Protection Policy

Because of increasing concerns about contamination of groundwater, the Authority has formulated a policy for groundwater protection which describes where discharge consents and prohibitions will operate as principal mechanisms for the control of polluting entries into groundwaters that are perceived to be at risk. Accordingly, the provision of oil interception, and other pollution prevention facilities are likely to become standard requirements where groundwater contamination is a hazard.

6.2.3 Adoption by sewerage undertakers

(a) Agreement with developer on sewers

Another important issue which may arise in the construction of an infiltration facility is that of eventual adoption of the facility by the appropriate sewerage undertaker. Ordinarily, a developer proposing to construct a sewer may enter into an agreement with the undertaker to the effect that, if the sewer is constructed in accordance with agreed specifications, it will be adopted by the undertaker. Thereafter, the undertaker will own the sewer and be responsible for its maintenance.

(b) Reasons for non-adoption of infiltration drainage systems

In relation to infiltration systems, however, some uncertainty surrounds the question whether a facility of this kind is within the legal definition of a 'sewer' and whether an undertaker would have grounds to decline to agree to adoption of such a facility purely because it comprised an infiltration, rather than a conventional, system of drainage. It is the policy of the Statutory Sewerage Undertakers in England and Wales not to adopt infiltration systems.

6.2.4 Acceptance of responsibility by highway authorities

Where an infiltration facility is proposed by a developer to serve as a highway drainage system, an analogous issue arises as to whether the facility may become the subsequent responsibility of the highway authority. In relation to this, a developer may enter into an agreement with a highway authority so that a proposed road becomes dedicated as a highway, maintainable at public expense, providing that it is constructed to an agreed specification. Amongst other matters, an agreement of this kind will make provision for the road to be properly drained. In this respect the statutory definition of a 'drain' explicitly includes reference to a 'soakaway', thereby making it apparent that a highway authority may not decline to undertake future responsibility for a road merely because an infiltration facility has been used for the drainage of surface water.

6.2.5 Adoption by land drainage bodies

Water quality is not the only issue in relation to the law concerning infiltration systems. Matters of water quantity may also need to be taken into consideration. Although a general common law right exists allowing private owners of land to conduct land drainage work, increasingly this activity has become the concern of public bodies entrusted with statutorily defined powers to conduct operations to alleviate flooding and improve drainage. Flood defence and land drainage powers are given to internal drainage boards and local authorities to undertake works for securing the effective drainage of their respective areas. These discretionary powers are fairly general in their formulation and allow any act 'required for the drainage of land'.

The implication of this is that, under appropriate circumstances, infiltration drainage may be adopted by these bodies if it constitutes an effective means of securing a drainage objective within their powers.

6.2.6 Maintenance responsibilities

(a) Single owner

Another consideration of potentially fundamental importance is that of continuing responsibilities for maintenance of infiltration facilities. This problem will be largely obviated where the particular facility is maintained by a sewerage undertaker or a highway authority, if necessary on terms which reflect the continuing maintenance costs involved, but otherwise the question of continuing responsibilities for maintenance of the facility will have to be resolved.

(b) Shared ownership

In a situation where more than one private individual or body possesses responsibility for maintenance of the facility, covenants would need to be entered into to provide for this. However, difficulties may arise in law in enforcing a burdensome covenant to subsequent owners of land concerned, and other legal mechanisms may need to be considered as a means of achieving the objective of securing continuing obligations in respect of maintenance. One legal mechanism which may secure the more effective imposition of continuing maintenance obligations is that of a rentcharge in respect of properties served by a communal infiltration facility, requiring each owner to pay an annual sum proportionate to the cost of maintaining the facility.

6.2.7 Civil liability and redress

(a) Extent of liability

Whether an infiltration facility is the continuing responsibility of a sewerage undertaker, highway authority, developer or a private individual, a vital question arises as to the extent of the civil liability which may occur where the operation of the facility results in harm to another individual or body. Here the essential complaint will be that the facility has brought about an adverse effect upon water quality or water quantity or has led to geotechnical problems which are so detrimental to the interests of another person that compensation should be payable or some other remedy given.

(b) Mechanisms for redress

The categories of legal action under which civil liability may be established are rather diverse, but of particular importance are negligence, breach of statutory duty, nuisance and liability under the principle established in the case of *Rylands v Fletcher*. These different forms of civil liability have distinct characteristics, but provide a legal mechanism by which redress may be made available to those who suffer a loss consequent upon the improper construction or maintenance of an infiltration facility. Accordingly, they must be taken into consideration by those with initial or continuing responsibilities for this kind of facility.

(c) Cambridge Water Company Case

In the case of *Cambridge Water Company v Eastern Counties Leather Plc*, the House of Lords reaffirmed that the leather company was not civilly liable for the losses incurred by the water company due to unforeseeable contamination of a water supply. The ruling emphasised that foreseeability of harm was an essential element in establishing civil liability in these circumstances.

Some care must be taken in the interpretation of the *Cambridge Water Case*. Notably the House of Lords was concerned not to impose an unreasonably high standard in relation to 'historic' pollution. Implicitly, it was held to be wrong to impose liability upon present day landowners for past activities, the polluting potential of which was not appreciated at the time they occurred. It would be no defence in a present day civil action, however, for a polluter to claim that he did not know, or could not foresee, the contaminating potential of a chemical spillage upon water to be used for supply purposes. The key point is that the foreseeability of harm must be judged at the time when the offending activity took place and not in the light of later knowledge of the adverse environmental effects of the activity.

Relating this back to the problem of pollution from infiltration drainage systems, if it is established that water which discharges into such systems may have a contaminating effect upon groundwater sources of supply, then the potential for civil liability will clearly exist. Care is needed in the design of such systems to ensure that no foreseeable contamination may take place. The absence of care in this respect may, as the House of Lords decision indicates, give rise to claims for compensation of considerable magnitude.

6.2.8 The European Community Groundwater Directive

Water pollution is not only a matter of national concern. It is equally provided for by way of European Community obligations under various Directives relating to water quality. Perhaps most notable among these is the European Community Groundwater Directive (no. 80/68/EEC) which seeks to prevent the pollution of groundwater by specified substances and to eliminate the consequences of existing pollution. Whilst, in most respects, this Directive requires discharges containing these substances made to groundwater to be subject to authorisations granted by the appropriate national authority, it also specifies that authorisations may only be granted where there is no risk of groundwater pollution. Accordingly, the existence, or otherwise, of a risk of that kind will be ascertained by the National Rivers Authority, in England and Wales, in consideration of the conditions to be applied where a consent to discharge is required.

6.3 ADMINISTRATIVE ASPECTS

The main administrative procedures involved in progressing infiltration drainage are outlined below.

6.3.1 Roles of the local authority

The local authority has many roles in drainage administration. It is the planning authority which examines the developer's planning application for strategic and technical compliance with the County Structure Plan, the Local Plan and the Unitary Development Plan. It may also act as an agent for the sewerage undertaker and the highway authority. It will usually grant or refuse planning permission in accordance with the drainage advice it receives from statutory consultees named in government circulars and in the interests of the public, commerce and the environment.

6.3.2 Planning controls

(a) Application for planning permission

Initially, the construction of an infiltration drainage facility will be a 'development' of land which requires planning permission from a local planning authority. Although no direct guidance is provided as to how the planning authority should determine an application for planning permission incorporating an infiltration drainage facility, past decisions have indicated that a material consideration in any planning determination will be the adequacy of the drainage system proposed and, the likelihood of flooding.

(b) Consideration by planning authority

The development control process requires the planning authority to examine all relevant drainage matters along with other considerations and implicitly, therefore, a planning authority will need to consider the suitability of a proposed infiltration system in relation to a development project before planning permission can be given.

(c) Technical evidence in support of application

It is likely that the fullest possible technical evidence will be required to support an application. It is recommended that this should include:

- size and nature of the area to be drained
- survey of ground conditions
- details of infiltration system to be used
- results of soakage test
- hydraulic calculations to demonstrate the adequacy of the system
- hydrogeological assessment of the risk of pollution of groundwater by stormwater runoff.

If Consent to Discharge or Land Drainage Consent is required the planning authority will normally have to be satisfied that these approvals have been obtained before they issue planning permission.

(d) Planning inquiry by DOE

A controversial proposal may be called in by the DOE and a planning inquiry held. An application which has been refused or delayed may also be reviewed by a planning inspector at a planning appeal. Drainage matters are usually discussed and technical proposals prepared ahead of a planning application so that all parties know what is planned and may respond when asked to comment.

(e) Accommodation of objections

A developer may offer to carry out necessary infrastructure work, including drainage for the site, using a formal Section 106 Agreement procedure specified in the Town and Country Planning Act 1990 to enable potential objections to be accommodated and planning permission granted. If this relates to work off-site, however, this may not always be practicable and in which case infiltration may beneficially form part of the solution. The developer may seek outline planning permission when the proposal is not complete and detailed permission when it is.

6.3.3 Building Control

(a) Building notice and plans

Further controls upon construction activities operate through the system of building control which, generally, requires any person who intends to carry out building work to provide the appropriate local authority or Approved Inspector with a building notice or full plans of the work which is proposed. Providing no defect is found in the plans the building control authority will approve them.

(b) Building regulations

Building regulations specify that systems which carry rainwater from the roof of a building to a soakaway or some other suitable rainwater outfall area must be adequate for these purposes. In other circumstances the use of an infiltration system will have to satisfy more general requirements relating to drainage to be acceptable.

(c) Evidence of suitability

The suitability of infiltration systems can be considered in a similar manner to that of a conventional soakaway. If the infiltration system requires any form of construction, therefore, it will need to satisfy building control. A building notice and full plans of the work will have to be submitted to the local authority or Approved Inspector. To obtain approval it is likely the fullest possible technical evidence will have to be presented, similar to that required for planning permission.

6.3.4 Consent to Discharge and Land Drainage Consent

If the discharge from the infiltration system is of a polluting nature then a Consent to Discharge must be obtained. If it incorporates an overflow to a watercourse then, in addition, it will require Land Drainage Consent.

(a) Consent to Discharge

The NRA is the appropriate regulatory body to authorise a Consent to Discharge and will require to know at least:

- the site of the proposed infiltration system
- the nature of the discharge, in particular its quality
- the expected volumes to be discharged.

Currently, the different regions of the NRA vary in their detailed practice but there is a progressive development of national guidelines that may be expected to produce future harmony. There are some differences in the organisation of the NRA regions but as a general rule reference should be made to the Area Planning Liaison Officer.

(b) Land Drainage Consent

Where an overflow to a watercourse is proposed and thus a Land Drainage Consent also applies, the NRA will be the consenting authority in cases where the watercourse is designated 'main river'; otherwise the approval of the relevant District Council would normally be required. In some low-lying areas Internal Drainage Boards exist and have purview over specific water courses for which their consent would be required.

7 Further issues

Preparation of this manual and its supporting reports has naturally highlighted a number of issues that need to be addressed. These are summarised below.

7.1 LEGISLATION

As is outlined in Section 6, there is little legislation that directly refers to infiltration systems and as a result, inferences have to be drawn by the application of broader principles to the specific issues raised by infiltration systems.

The Survey of Practice indicated that at present the lack of clear legislation referring to infiltration systems inhibits their use. The situation could be clarifed by legislation that includes explicit reference to infiltration systems and in particular addresses their eligibility for adoption.

7.2 ADOPTION

A particular issue which needs addressing relates to the ability of sewerage undertakers to adopt infiltration systems. The alleged requirement that a sewer has to have an outfall precludes infiltration systems from being considered as part of a public sewerage system. This has led to the current refusal of sewerage undertakers to adopt infiltration systems. This alleged ineligibility for adoption inhibits the use of infiltration systems where the developer wishes to have the sewerage system adopted and restricts the scope of an integrated approach to storm-water management.

7.3 LONG-TERM PERFORMANCE AND MAINTENANCE

A concern that is quite properly raised about infiltration systems is that their performance may deteriorate over time. There is little information on how infiltration systems behave in the longer term or on the impact that various levels of maintenance may have on longer-term performance. Information on the impact of ageing on the hydraulic performance of infiltration systems would be invaluable for improving confidence in the hydraulic design. At present there is little information on which to base the specification of appropriate factors of safety.

Though some types of infiltration system are relatively new to the UK, there is a long history of the use of soakaways. There should, therefore, be available a wide range of soakaways of different ages and which may have been subjected to a range of maintenance regimes. By investigating the performance of these and by carrying out hydraulic tests alongside them, it should be possible to determine long-term performance of such systems and the impact of different maintenance procedures.

7.4 MAINTENANCE PROCEDURES

Procedures for cost-effective maintenance also need to be developed. Clearly the problem is most acute for sub-surface systems and innovative thought needs to be given at the design stage to ways in which the maintenance of sub-surface systems can be improved.

7.5 WATER QUALITY

Frequently major concerns are expressed about the possibility of infiltration systems introducing polluted water into the groundwater system. The data presently available suggests that for many types of infiltration systems a high proportion of particulate pollutants is retained within or close to the infiltration system. There is, however, not enough information available at the moment to provide a scientific method for quantitative estimates of the proportion of contaminants that would be released to groundwater sources.

Thus experiments and fieldwork should concentrate on determining the percentage of incoming pollutants that are removed, either by physio-chemical or biological means within the unsaturated zone. The concern is both the long-term removal rate and also the amount by which the peak concentration is reduced during a single storm event.

Of particular concern are soluble pollutants such as TBME which cannot feasibly be separated from water using either physical or biological methods.

7.6 ANALYSIS OF THE RISK OF GROUNDWATER POLLUTION

There is a risk that surface water runoff may be polluted, either through the nature of the catchment use, carelessness, or as a result of an accident and that this could lead to contamination of groundwater. The difficulty of dealing with polluted aquifers means that the prevention of such pollution is essential to preserve the quality of the groundwater resource. In order to do this it is important to determine the magnitude of this risk.

A procedural framework is required for carrying out a risk analysis that will identify the magnitude of the threat to groundwater quality. Such a framework would include a hydrogeological assessment of the impact of potential pollution from infiltration systems on groundwater quality and abstractions.

Such a framework should address the range of hazards which can result in pollutants entering an infiltration system, through either normal or accidental usage and the likely frequency of each. For each hazard, guidance needs to be formulated on assessing the potential risk of pollution of the groundwater. The aim would be to increase the use of infiltration systems with the minimum risk of contaminating the aquifer.

7.7 COST DATA

There is a need for more cost data, particularly on the operation, maintenance and replacement costs of infiltration systems. These should be based on a level of operation and maintenance to provide satisfactory service rather than on the random basis of crisis management. However, as a starting point, the actual costs of maintenance should be established.

8 Conclusions

1. The purpose of this manual is to promote the use of infiltration systems as one of a range of possible options for controlling and disposing of stormwater runoff. The range of options available to effect infiltration of stormwater runoff at source has been set out in this manual, and provides the designer with the tools to assess the viability of such systems in particular localities and circumstances.

2. The range of applications can be seen from the series of case studies described in the associated publication, CIRIA Project Report 22, which range from soakaways through swales, trenches, basins and pavements. All proved the potential for successful application.

3. There is no doubt that the use of infiltration systems reduces the impact on the drainage system and watercourses downstream of increases in runoff due to urbanisation. In consequence, the capacity required of the piped system is reduced and there is a lower incidence of stormwater overflow discharges polluting watercourses.

4. Significant increases can also be obtained in the quantity of groundwater recharge. It must be noted, however, that there is a risk of pollutants being introduced into the soil and groundwater, although there are methods available for reducing these risks.

5. There is a need for regular maintenance and proper arrangements must be made for carrying this out.

References

Bettess R, Davis A, and Watkins D C, 1995, *Infiltration Drainage – Hydraulic Design*, CIRIA Project Report 23

Building Research Establishment, 1973, *Soakaways*, BRE Digest 151, BRE, Watford

Building Research Establishment, 1991, *Soakaway design*, BRE Digest 365, BRE, Watford

British Standard Institute, 1985, BS 8301 – *Code of Practice for Building Drainage*

Cedergen HR, 1974, *Drainage of highway and airfield pavements*, J Wiley

CIRIA, 1992a, *Scope for control of urban runoff:* Volume 1, CIRIA Report RP123

CIRIA, 1992b, *Scope for control of urban runoff:* Volumes 2, 3 and 4, CIRIA Report RP124

CIRIA, 1994, *Control of pollution from highway drainage discharges*, CIRIA Funders Report IP/3 and CIRIA Report 142

CIRIA 1995, *Infiltration Drainage – Literature review*, CIRIA Project Report 21

CIRIA 1995, *Infiltration Drainage – Case studies of UK practice*, CIRIA Project Report 22

CIRIA 1995, *Infiltration Drainage – Legal aspects*, CIRIA Project Report 25

Colwill D M, et al 1984, *Water quality of motorway runoff*, Transport and Road Research Laboratory. Supplementary Report 823

Coppin N J and Bradshaw A D, 1982, *Quarry reclamation – the establishment of vegetation in quarries and open pit non-metal mines*, Mining Journal Books Ltd, London

Coppin N J and Richards I G, 1990, *Use of vegetation in Civil Engineering*, Butterworths, London

Diniz E V, 1980, *Porous Pavement Phase 1 – Design and Operation Criteria*, US EPA, Report No EPA-600/2-80-135

Ellis J B, 1989, *The development of environmental criteria for urban detention pond design in the UK* in Design of Urban Runoff Quality Controls. Proc. of an Engineering Foundation Conference, Missouri, July 1988, edited by Roesner LA, Urbonas B and Sonnen MB, pp 14-28

Ellis J B, 1991, *The design and operation of vegetation systems for urban runoff quality control*, Proc of 3rd Standing Conference on Stormwater Science Control

Ellis J B, 1992, *Design criteria for managing detention basin quality* in Urban Stormwater Management edited by O'Loughlin G, Australian Institute of Engineers

Escritt L.B. and Haworth W.D. (1984). *Sewerage and Sewage Treatment – International Practice*

HM Treasury, 1984, *Investment appraisal in the public sector:* A technical guide for Government Departments

Hogland W, Niemczynowicz J and Wahlman T, 1987, The unit superstructure during the construction period, *The Science of the Total Environment,* No 59, pp 411-424

Hogland W, 1990, *Pervious asphalt construction. An overview of the situation in Sweden and the United States* in *Infiltration and storage of stormwater in New Development* edited by Balmforth D, pp 122-135

Hubbard C E, 1986, *Grasses* (2nd edition), Penguin, London

Institute of Hydrology, 1975, *Flood Studies Report,* Volumes 1 to 5, NERC, London

National Rivers Authority, 1992, *Policy and Practice for the Protection of Groundwater,* NRA

Pratt C J, 1989, *Permeable pavements for stormwater quality enhancement,* ASCE Engineering Foundation Conference, Urban Stormwater Quality Enhancement – Source control, retrofitting and combined sewer technology, Daros, Switzerland, October

Price M, 1994, *Drainage from roads and airfields. To soakaways: Groundwater pollutant or valuable recharge?* Journal of the Institution of Water and Environmental Management, 8, (5): pp 468-479

Stahre P and Urbonas B R, 1990, *Stormwater detention,* Prentice Hall

Strecker E W et al, 1990, The US Federal Highways Administration receiving water impact methodology. *The Science of the Total Environment No 93,* pp 489-498

Subsidence Engineers Handbook, 1975

Watkins D C, 1991, *The hydraulic design and performance of soakaways,* HR Wallingford Report SR271, HR Wallingford

Watkins D C, 1995, *Infiltration drainage – Literature Review,* CIRIA Project Report 21

Appendix Relationship between the method of hydraulic calculations in CIRIA Report 124 (Volume 3), BRE Digest 365 and the method presented in this manual

The design assumptions made in the various documents are different and this can result in different designs. A comparison of the main features assumed in each document are given below:

Size of trial pit

CIRIA's Report 156 recommends that the size of the trial pit should be related to the area to be drained with a volume of at least 0.5 m^3 for an infiltration system to drain 100 m^2 and a volume of at least 2.0 m^3 if the area to be drained is greater than 100 m^2.

BRE Digest 365 recommends that the test pit should be 0.3 to 1 m wide, 1 to 3 m long and to the same depth as that anticipated for the soakaway. This will, in general, imply a larger test pit than that recommended in FR/IP/5.

CIRIA Report 124 is not specific about the size of test pit required, but it would appear to imply pits larger than those recommended in FR/IP/5.

As infiltration systems can be used to drain areas of very different sizes, it seemed reasonable to allow a degree of flexibility in selecting the size of the trial pit.

Hydraulic calculations

In CIRIA Report 124 and BRE Digest 365 it is assumed that the infiltration from the system into the soil is constant and corresponds to that when the system is half full of water. In FR/IP/5 the infiltration is assumed to depend upon the wetted area and so varies with the depth of water in the system. For shallow depths the method in FR/IP/5 assumes a smaller outflow from the system than CIRIA Report 124 and BRE Digest 365, while for larger depths the method in FR/IP/5 assumes a larger outflow.

The hydraulic calculations were based on work carried out at HR Wallingford (Watkins, 1991) and seem to provide a more realistic representation of the hydraulics of infiltration systems. Unfortunately, the result is that the hydraulic calculations are more complex than those in CIRIA Report 124 or BRE Digest 365 but can still be performed on a calculator or using a computer spreadsheet.

Infiltration area

CIRIA Report 124 and BRE Digest 365 assume that no infiltration takes place through the base of the system, while FR/IP/5 assumes that it does.

For plane infiltration systems all the infiltration takes place through the base of the system and so an assumption of no flow through the base is impractical.

Factor of safety

There is no factor of safety included in CIRIA Report 124 or BRE Digest 365 for the consequences of progressive siltation. FR/IP/5 uses a factor of safety which depends upon the consequences of failure.

In view of the uncertainties associated with both rainfall statistics and the hydraulic predictions, it was considered to be prudent to include a factor of safety. This was thought to be particularly appropriate where large infiltration systems were being considered. For small systems draining small areas the volume of flooding and the resulting consequences are normally small. For larger systems the volume of flooding could be very much larger and the consequences correspondingly more severe.